Animal Sonar Systems

NATO ADVANCED STUDY INSTITUTES SERIES

A series of edited volumes comprising multifaceted studies of contemporary scientific issues by some of the best scientific minds in the world, assembled in cooperation with NATO Scientific Affairs Division.

Series A: Life Sciences

Recent Volumes in this Series

The series is published by an international board of publishers in conjunction with NATO Scientific Affairs Division

A Life Sciences	Plenum Publishing Corporation
B Physics	London and New York
C Mathematical and Physical Sciences	D. Reidel Publishing Company Dordrecht, Boston and London
D Behavioral and Social Sciences	Sijthoff & Noordhoff International Publishers
E Applied Sciences	Alphen aan den Rijn, The Netherlands, and Germantown U.S.A.

Animal Sonar Systems

Edited by
René-Guy Busnel
Ecole Pratique des Hautes Etudes
Jouy-en-Josas, France

and

James F. Fish
Sonatech, Inc.
Goleta, California

SPRINGER SCIENCE+BUSINESS MEDIA, LLC

Library of Congress Cataloging in Publication Data

International Interdisciplinary Symposium on Animal Sonar Systems, 2d, Jersey, 1979.
Animal sonar systems.

(NATO advanced study institutes series: Series A, Life sciences; v. 28)
Symposium sponsored by the North Atlantic Treaty Organization and others.
Bibliography: p.
Includes indexes.
1. Echolocation (Physiology) – Congresses. I. Busnel, René Guy. II. Fish, James F.
III. North Atlantic Treaty Organization. IV. Title. V. Series.
QP469.I57 1979   599'.01'88   79-23074
ISBN 978-1-4684-7256-1   ISBN 978-1-4684-7254-7 (eBook)
DOI 10.1007/978-1-4684-7254-7

This work relates to Department of the Navy Grant N00014-79-G-0006
issued by the Office of Naval Research. The United States Government
has a royalty-free license throughout the world in all copyrightable
material contained herein.

Proceedings of the Second International Interdisciplinary Symposium on
Animal Sonar Systems, held in Jersey, Channel Islands, April 1–8, 1979.

© 1980 Springer Science+Business Media New York
Originally published by Plenum Press, New York in 1980
Softcover reprint of the hardcover 1st edition 1980
A Division of Plenum Publishing Corporation
227 West 17th Street, New York, N.Y. 10011

ORGANIZING COMMITTEE

- R.G. Busnel
  Laboratoire d'Acoustique Animale
  E.P.H.E. - I.N.R.A. - C.N.R.Z.
  78350 Jouy-en-Josas, France

- J.F. Fish
  U.S. Navy
  Naval Ocean Systems Center
  Kailua, Hawaii 96734, U.S.A.

- G. Neuweiler
  Department of Zoology
  University of Frankfurt
  D-6000 Frankfurt, F.R.Germany

- J.A. Simmons
  Department of Psychology
  Washington University
  St. Louis, Mo. 63130, U.S.A.

- H.E. Von Gierke
  Aerospace Medical Research Lab.
  Wright Patterson Air Force Base
  Dayton, Ohio 45433, U.S.A.

A C K N O W L E D G M E N T S

-=-=-=-=-=-=-=-=-=-=-=-=-=-=-

This symposium was sponsored by various Organizations, National and International. The organizing Committee would like to thank them and their representatives :

- North Atlantic Treaty Organization, N.A.T.O.

- Advanced Study Institutes Programme

- Office of Naval Research of the U.S.A.

- United States Air Force

- Volkswagen-Stiftung from F.R. of Germany

- Ecole Pratique des Hautes-Etudes, Laboratoire d'Acoustique Animale, I.N.R.A. - C.N.R.Z. - 78350 JOUY-en-JOSAS, France

Preface

Thirteen years have gone by since the first international meeting on Animal Sonar Systems was held in Frascati, Italy, in 1966. Since that time, almost 900 papers have been published on its theme. The first symposium was vital as it was the starting point for new research lines whose goal was to design and develop technological systems with properties approaching optimal biological systems.

There have been highly significant developments since then in all domains related to biological sonar systems and in their applications to the engineering field. The time had therefore come for a multidisciplinary integration of the information gathered, not only on the evolution of systems used in animal echolocation, but on systems theory, behavior and neurobiology, signal-to-noise ratio, masking, signal processing, and measures observed in certain species against animal sonar systems.

Modern electronics technology and systems theory which have been developed only since 1974 now allow designing sophisticated sonar and radar systems applying principles derived from biological systems. At the time of the Frascati meeting, integrated circuits and technologies exploiting computer science were not well enough developed to yield advantages now possible through use of real-time analysis, leading to, among other things, a definition of target temporal characteristics, as biological sonar systems are able to do.

All of these new technical developments necessitate close cooperation between engineers and biologists within the framework of new experiments which have been designed, particularly in the past five years.

The scientists who have been working on these problems in various fields (electronics experts, signal processors, biologists, physiologists, psychologists) have produced new and original results, and this second symposium furnished the opportunity of cross-disciplinary contacts permitting an evaluation of the state of present research.

The Jersey meeting in April, 1979, brought together more than 70 participants from 8 different countries. This meeting was particularly necessary considering the number of new research groups that have appeared in various fields. I mention in particular: the Federal Republic of Germany, where two young scientists who participated at the Frascati meeting, Gerhard Neuweiler and Hans-Ulrich Schnitzler, have since become professors and have founded two schools of highly productive research; the United States, where funded by the U.S. Navy, studies on dolphins have spread to San Diego and Hawaii, given impetus by Bill Powell, Forrest Wood, Sam Ridgway, C. Scott Johnson, Bill Evans, and Ron Schusterman, and where certain of Donald Griffin's most gifted students, such as Alan Grinnell and Jim Simmons, are continuing his work on bats at various universities; Canada, where Brock Fenton is performing outstanding research.

Although much research has been carried out in the Soviet Union since 1969-1970, it is most unfortunate that, for reasons independent of their wishes, our colleagues from this country, who have moreover published several excellent reviews of their work, were not able to participate in our discussions.

The tendancy which appeared at Frascati towards a certain zoological isolation corresponding to a form of animal specialization dominated by either dolphins or bats, has partially regressed, thanks to several physicists who use both bat and dolphin signals in their theoretical approaches. While this segregation by field still remains a dominant behavior made obviously necessary, up to a certain point, by the different biological natures of these two groups of animals, the phenomenon is aggravated by the use of semantics specific to each group. Nevertheless, common interests and attempts at mutual understanding which appeared are encouraging and should be congratulated.

The proceedings of the Jersey meeting demonstrate the extent to which technology has advanced in the past decade, in performances of transducers, various microphones, hydrophones, as well as in recording apparata, analytical methods, particularly in the use of signal processing techniques, and in the application of new ideas such as time-domain, auditory processing, frequency-domain, Doppler compensation, target-acoustic imaging, and so on.

It is also interesting to note to what extent experimental strategies have been refined, and one can only admire the elegance of certain demonstrations carried out on dolphins as well as on bats. Many aspects of the performances of diverse species continue, however, to intrigue us, as they reveal sensory abilities whose fine analysis still eludes us, particularly the central mechanisms which control and regulate them. As an example of this, I would particularly like to mention how enriching was the experience that we were able to have using Leslie Kay's apparatus for the blind,

which gave spatiodirectional sensations analogous to those of air-
borne animal sonar systems. I do not doubt in the least that those
several minutes will lead to a totally new concept of biosonar problems.

As Henning von Gierke pointed out during the last session and
in a personal communication, the trends appearing at the Jersey
meeting indicate that pattern recognition theory is becoming more
and more important to biosonar research, replacing range finding
and echo theories as the promising research areas of the future.
Pattern recognition specialists should be included in future meet-
ings as well as experts on the spatial frequency analysis of acoustic
and visual perception. Although animal sonar might be used to a
large extent for acoustic imaging of space, we know from ultrasono-
graphy that acoustic images are different from optical images.
Acoustic space perception therefore differs from visual space perception. Since the acoustic space is scanned sequentially, the
total "acoustic image" depends primarily on memory capability, which
is the major difficulty encountered by Leslie Kay in his device for
the blind. The findings of this Symposium may have a major impact
on general auditory physiology regarding, on one hand, the example
of sharp filter (acoustic fovea) and on the other, peripheral processing.

A reading of the present set of volumes presenting the current
state of research will bring out at the same time the unknowns of
the problem, the uncertainties, the hypotheses, and will allow vet-
erans of Frascati to measure the progress made since then.

I am most happy to thank here my American colleagues who agreed
to respond to my call in 1977, particularly Henning von Gierke, who
was our referee and support for N.A.T.O. as well as the U.S. Air
Force, and Bill Powell and Forrest Wood for the U.S. Navy, who later
recommended to me Jim Fish, a young associate, dynamic and efficient,
who held a preponderant place in the Organizing Committee.

I also would particularly like to express my gratitude to Gerhard
Neuweiler for his constructive participation in our planning group.
It is thanks to his outstanding reputation that our Symposium was
able to obtain funding from Volkswagen for the active and highly
productive German delegation. He assumed as well the heavy respon-
sibility of financial management of our funds.

I also wish to thank, on behalf of our Committee and myself,
the diverse individuals from my laboratory who, with devotedness,
took on countless tasks, often thankless and lowly: Michèle Bihouée,
Annick Brézault, Marie-Claire Busnel, Sophie Duclos, Diana Reiss,
and Sylvie Venla. Leslie Wheeler, who assisted me in editing these
volumes, deserves special mention, as without her help I would not
have been able to publish them so rapidly.

During the last plenary session in Jersey, the Organizing
Committee and the Co-chairmen of the different sessions decided on
the publication format of the proceedings, and suggested to the
Symposium participants to dedicate this book to Donald R. Griffin.
Our colleagues unanimously rendered homage to the spiritual father,
the inventor, of echolocation.

The scientific wealth brought out during the three half-day
poster sessions, bears witness to the interest and importance of
the work of numerous young scientists and makes me optimistic for
the future of the field of animal sonar systems. For this reason,
and thanks to the two experiences which the majority of you have
considered successful, I wish good luck to the future organizer of
the Symposium of the next millenium.

René-Guy Busnel
Jouy-en-Josas
France

Contents

On behalf of the members of the organizing Committee and all the Symposium participants, we dedicate this book to

Donald R.GRIFFIN

in homage to his pioneering work in the field of echolocation.

Donald R. Griffin

DEDICATION

by Alan D. Grinnell

Just over 40 years ago, a Harvard undergraduate persuaded a physics professor to train his crystal receiver and parabolic horn at active bats. They detected ultrasonic pulses, and the contemporary field of echolocation research was born. The undergraduate was Donald R. Griffin, the professor, G. W. Pierce, and their first report appeared in 1938. In recognition of his founding role and his countless important contributions to echolocation research this volume is dedicated to Don Griffin. Fortuitously, this coincides with the approach of his 65th birthday, a time when particularly popular and influential figures in a field are often honored with a "Festschrift" volume.

We were fortunate, at this meeting, that Don Griffin was persuaded to add a few recollections of his experiences during the infancy of the field. These are included in this volume. Additional perspective on some of the early years has been volunteered by his partner in the first experiments demonstrating echolocation, Dr. Robert Galambos, then a Harvard graduate student, now a well-known auditory neurophysiologist and Professor of Neuroscience at UCSD.

"In early 1939 Don found out that Hallowell Davis at the Harvard Medical School was teaching me how to record electric responses from guinea pig cochleas and asked if I could slip in a few bats on the side. He wanted to know whether their ears responded to the "supersonic notes" he and Professor G. W. Pierce had just discovered. So I asked Dr. Davis, who said "go ahead" and thus Don found himself a collaborator in some unforgettable adventures. (This was not the only time Don enlisted me in one of his enterprises--I remember spending 3 days helping him build a bird blind on Penekeese Island in weather so foul the Coast Guard had to come rescue us.)

During that spring I worked out the high frequency responsivity of the bat cochlea, and in the fall we assessed the obstacle avoidance capabilities of the 4 common New England species (using an array of wires hanging from the ceiling to divide a laboratory room into halves). We recorded the inaudible cries bats make in flight and demonstrated (by inserting earplugs or tying the mouth shut) that they must both produce and perceive them if their obstacle avoidance is to be successful. (We also made a sound movie of all this which nobody can find.) That research yielded my Ph.D. thesis and launched Don on the career this book honors.

All the crucial new measurements we made used the unique instruments devised by Professor Pierce. A physics professor who, like Don, seemed to me really a naturalist at heart, Pierce had designed his "supersonic" receivers in order to listen to the insects singing around his summer place in Vermont (or was it New Hampshire?). How Don found Professor Pierce and then talked him into letting us use his apparatus I do not know.

In the spring of 1940, when we felt we understood how bats avoided obstacles in the laboratory, it occurred to us to test our ideas in the field. So we made an expedition to a cave Don knew in New York State taking along the portable version of Pierce's supersonic receiver shown in the accompanying picture (Don was a skillful photographer even in those days).

The cave in question opened just beyond the bank of a mountain brook. Once inside you first climbed down a ways and then up. We had no trouble getting in and, after the small upward climb, found ourselves looking along a straight tunnel as far as our flashlights could penetrate. We set up the equipment and Don went further on to the gallery where the bats roosted. He sent several of them flying, one by one, down the tunnel in my direction and as they approached Professor Pierce's machine emitted the chattering clamor we hoped for. Then Don and I changed places so he could hear the noise too. Since the readout of Pierce's portable device came only via earphones, the sole record of the first bat ultrasonic cries ever heard outside a laboratory is the one engraved in our memories.

Throughout all of this Don kept urging me not to waste time, and I thought he closed off the exercise and moved us toward the entrance after an almost indecently short interval. He explained his behavior once we had scrambled down and then up and then out to cross the brook: the spring sun was rapidly melting the snow, the brook was rising fast, and he had been worried from the moment we arrived that the water might fill

the entrance to the cave and trap us inside. My captain, I
thought to myself, always looks out for the safety of his crew."

.During the nearly four decades since that first demonstration
of echolocation, the field has flourished. This is evident from
the quality and diversity of work described in this volume. It can
also be seen, quantitatively, in the accompanying graph, which shows
the number of publications on echolocation in microchiropteran bats
during each 3-year period since 1938. This graph, kindly prepared
by Uli Schnitzler, shows several surges in productivity - one about
1961, when the second generation of U.S. scientists began publishing,
and a very large one in 1967, when, following Frascati, the second
generation of German scientists and the Russian group entered the
field. Now, with the Jersey Symposium, the third generation is beginning to contribute importantly. Throughout this entire period,
while moving professionally from Harvard to Cornell, back to Harvard,
and then to the Rockefeller University, and while making major contributions to a number of other fields, Don Griffin has continued to
guide development of the field of echolocation with imaginative, incisive experiments: demonstrating the usefulness of echoes for orientation in the lab and in the field, showing that echolocation was
used for insect capture, documenting the sensitivity of the system
and its resistance to jamming, developing techniques to show how
accurately targets can be discriminated, and analyzing the laryngeal

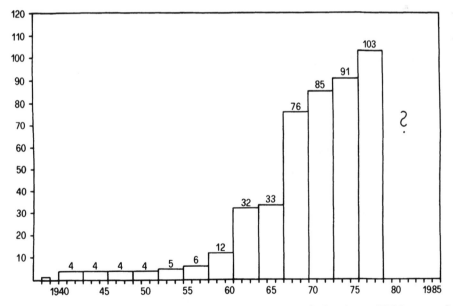

Histogram of papers published per 3-year period since 1938 on echolocation in microchiropteran bats.

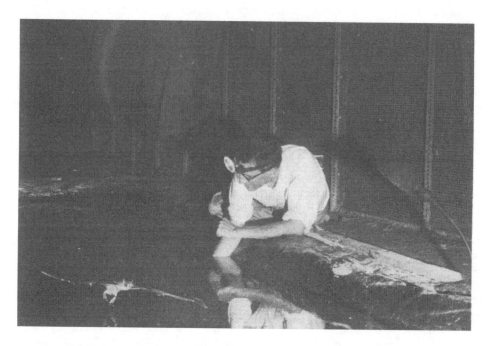

D. R. Griffin at Simla, Trinidad, studying capture of food by
<u>Noctilio</u> (1960).

---

mechanisms involved in ultrasonic orientation. It was in his lab,
with his encouragement, that the first major advances were made in
understanding the neural adaptations for echolocation, an aspect of
the field that has grown to enormous and impressive proportions. In
association with Al Novick, he established the great value of comparative approaches to echolocation, traveling the world not only to
document the variety of sounds and skills shown by bats, but to do
the first careful studies of echolocating birds, as well. Out of
early interests in the feeding habits, home ranges, and seasonal mi-
grations of bats (not to mention a long-standing fascination with
homing and migration in birds) came pilot studies on the role of
echolocation, passive hearing and vision in homing behavior. Indeed,
in countless instances, he has broken new ground and set standards
of rigor and experimental elegance that have served the field well.

A major milestone in the field of animal behavior was the publication, in 1958, of <u>Listening in the Dark</u>, Griffin's monograph on
his experiments and thoughts on echolocation, and winner of The Daniel
Giraud Elliot Medal of the National Academy of Sciences. With the
perspective of 20 years subsequent work in the field, it is aston-
ishing to anyone re-reading this classic to realize how fully Griffin
already understood the phenomenon of echolocation, how many critical
discoveries he had already made, and how profound were his insights.
This was one of five books Don Griffin has written, the most recent

of which is "The Question of Animal Awareness: Evolutionary Con-
tinuity of Mental Experience" (1976).

It was not only for his incisive early work that Don Griffin
deserves recognition as "father" of the field (or Godfather as he
was also described during the meeting); he has also been the academic
father, or colleague, of a high percentage of those who have become
active in the field. Most of the contributors to this volume who
have worked on bats have felt the imprint of Griffin's personality
and experimental approach directly, as graduate students or post-
doctoral research associates working in his laboratory, or as their
students, and it has been a powerful influence on their lives and
careers.

Don is a great storyteller, as his chapter in this volume at-
tests. More than that, however, his approach to science is guar-
anteed to never leave a dull moment. Whether it is fighting to stay
aloft after releasing birds from a small plane piloted by Alex Forbes,
or trying to convince suspicious authorities that he had valid reason
for clamboring around the fire escapes of a mental hospital, or barely
escaping a flooded cave, or covering for a student who had fallen
through the ceiling of the Mashpee Church, or searching the tombs of
an ancient Italian cemetery for Rhinolophus, his approach to experimentation is as direct and audacious as his ideas. Many of the
breakthroughs in echolocation research were the result of his intro-
duction of new or unfamiliar technology, from home-made ultrasonic
microphones to high speed tape recorders, sonagraphs, and information
theory. This applies in all of his other fields, as well. In recent
years, for example, much of his energy has been devoted to radar
identification and tracking of migrating birds. One of his collab-
orators in this research, Charles Walcott, tells of just one of the
complications that this has gotten them into, and of Don's characteristic aplomb in solving the problem:

"That season Don arrived in Stony Brook with a trailer
carrying what looked like a giant coffin. Actually the box
contained a helium filled kitoon - a sort of hybrid balloon -
kite combination that could be flown up to several thousand
feet in the air. Its purpose was to place instruments in the
same air mass that the radar showed birds to be flying in.
Indeed on its trials in Stony Brook the kitoon worked splen-
didly - we were able to get it up at least a thousand feet
or more - amply high enough to be where birds flew. Unfor-
tunately they aren't the only things that fly there - a
helicopter soon appeared flying significantly lower than the
kitoon. As it came close and inspected the kitoon we noticed
that the helicopter bore the inscription "Police" in large
letters. We rapidly hauled down the kitoon and anxiously
awaited the arrival of sirens and patrol cars; fortunately
they never came. But on other occasions, the law has become

involved and Don had been read the Federal Air Regulations
which prohibit the flying of balloons, kites or other such
devices more than a few feet above ground. The State Police
in Millbrook, New York next to The Rockefeller field station
had alerted the FAA about Don's activity. Don promised the
FAA representative that it wouldn't happen again. It was the
following Easter Sunday when Don and his colleagues were
following a kitoon with radar that the kitoon's string broke.
Following the kitoons progress with the radar, they saw that
it suddenly stopped drifting in the wind and became stationary.
It was hanging only a few hundred feet in the air directly over
the State Police headquarters. The string had caught in the
upper branches of a small sapling. With a typical display of
ingenuity, Don rigged a second kitoon with a grappling iron
and managed to snag the string of the escaped kitoon retrieving
both just before dawn on Easter Sunday!"

To those of us who have had the privilege of working with Don
Griffin, perhaps the greatest lesson has been his emphasis on rigor
in experimentation. No one is more acutely aware of the ambiguities
of an experiment, the difficulties involved in demonstrating something convincingly. Nor does he ever jump to conclusions. His in-
tellectual vigilance has been characterized by the claim that if he
were in a car passing a flock of sheep in a field, and a travelling
companion commented on the fact that among the sheep were two that
were black, he would reply, "They're black on the side facing us,
anyway." (Attributed to Don Kennedy.) On the other hand, he is not
afraid to consider some of the most complex formsof animal behavior,
and to take on established dogma, decrying the excessive use of "sim-
plicity filters" in interpreting behavior, a position he argues elo-
quently in his recent writing on animal awareness.

This mixture of insistence on careful unbiased observation of
what is really there and rigorous proof of any conclusions, combined
with a brilliant imagination and willingness to adapt new technology
to biological problems, has done much to develop the field of animal
behavior research. It has served as particular inspiration for all
of us.

We are honored to dedicate this volume to Donald R. Griffin.

AN ANALOGUE DEVICE FOR THE GENERATION OF SONAR AMBIGUITY DIAGRAMS

Justin A. T. Halls

The ambiguity diagram is a measure of the sonar or radar properties of a given echolocation signal. It consists of a three-dimensional graph in which the x and y axes represent range and velocity, and the height at any point is the probability of a target having the corresponding range/velocity co-ordinates. Each section through the diagram parallel to the range axis represents the output of an ideal, range measuring receiver (eg. matched filter or correlation receiver) in response to an undistorted, energy normalised echo from a target moving at the velocity corresponding to that section; this output is the cross-correlation function of the transmitted signal and a doppler shifted copy of the transmitted signal. Each section through the diagram parallel to the velocity axis represents the output of an ideal velocity measuring receiver (eg. Fourier Transform receiver) in response to an undistorted, energy normalised echo from a target at the range corresponding to that section.

The ambiguity function was first defined by Woodward (1953) as:-

$$|\chi(\tau,\phi)|^2 = |\int u(t) . u*(t+\tau) . e^{j\phi t} dt|^2$$

in which u(t) is the transmitted signal; τ = negative of time delay; $\phi = -(2\omega_0 v/c)$ = frequency shift caused by the Doppler effect; ω_0 is the carrier frequency in radians; c is the speed of signal propagation; v is the radial velocity of the target, taken to be positive for motion towards the receiver and negative away from it. This form of the ambiguity function is referred to as the narrowband approximation, since it was formulated for use with radar signals; it assumes that the bandwidth of the signal is small compared to the carrier frequency, and that the velocity of the target is small compared to the speed of propagation of the signal; the doppler effect can then be considered to be a simple spectral shift or frequency offset. In this case the height of the function at the origin, and the volume of the function are fixed and proportional to the energy contained in the original signal. Since the diagram is antisymmetrical about the origin it is only necessary to calculate half the function, the other half being a mirror image of this.

For wideband sonar signals the generalised ambiguity function must be used, although for convenience χ may be plotted rather than $|\chi|^2$.

$$|\chi(\tau,s)|^2 = |s^{1/2} \int u(t) . u*(s(t+\tau)) dt|^2$$

where s is the doppler stretch or compression factor and equals

$(1+\beta)/(1-\beta)$, and $\beta = v/c$. $s^{1/2}$ normalises the energy in the 'echo'.
This is the form of the ambiguity function used by Cahlander (1967)
to generate ambiguity diagrams of a number of echolocation signals
from Myotis lucifugus and Lasiurus borealis. In this wideband case,
the ambiguity diagram is not necessarily antisymmetrical, and its
volume is not fixed.

The ambiguity function computer described here uses analogue
techniques, and is being used to generate the ambiguity diagrams of
real bat signals from over 100 species, recorded under a variety of
conditions, by J. D. Pye. $\chi(\tau,s)$ is calculated for 1000 successive
values of τ, and the resulting correlation function is plotted.
This process is repeated for as many values of s as are required.
Up to 16 values of s, which may be +ve or −ve, may be calculated
in one run, although for most signals it is only necessary to
compute 5 different values.

The signal is stored at 1/8 or 1/10 speed on the rotating
magnetic drum of a Kay sonagraph, and is replayed continually into
the computer. The computer directs one copy of the signal into an
analogue delay line consisting of charge coupled analogue shift
registers. A second copy of the signal goes to a similar delay
line in which the delay varies continuously throughout the duration
of the signal. This results in a stretching or compression of the
signal in time, thus imposing a doppler shift proportional to the
rate of change of delay. The outputs of the two delay lines are
then filtered and multiplied in a balanced mixer. The product is
integrated and the result stored in the memory of a Biomac 1000
averaging computer. The delay in the first delay line is then
incremented and the process is repeated until a full 1000 point
correlogram has been generated. This is then plotted on an x-y or
y-t plotter, the doppler factor s is incremented, and the process
is repeated. For convenience the correlation functions themselves
are then assembled into the full ambiguity diagram, although a
squaring circuit and envelope detector may be inserted between the
Biomac 1000 and the plotter if desired.

The computer can also be used for generating cross-correlation
functions or cross-ambiguity functions between two different signals,
or between a signal and a real echo of that signal. A simple
modification also permits the generation of range/acceleration as
well as range/velocity ambiguity diagrams.

It should not be assumed that the bat is making use of all the
information that the ambiguity diagram indicates is available from
a given signal. The ambiguity function represents an upper bound
on the precision with which certain target parameters can be
measured, with a single pulse. In a real system the available
information will always be less than this. Also the bat is not

relying on a single pulse to provide a complete picture of his
environment, and many improvements can be made by using pulse
trains and by making a priori assumptions about the nature of the
target.   However, the ambiguity function does tell us something
of the sonar properties of a given pulse, enabling us to estimate
the type of system in which it would be of most use. As such the
ambiguity diagram is an essential supplement to the oscillogram
and the sonagram when attempting to describe signals. The computer
described here provides an economical way of producing ambiguity
diagrams for real or simulated pulses on a routine basis.

REFERENCES

Cahlander, D.A., 1967, Discussion of Theory of Sonar Systems, in
      "Animal Sonar Systems: Biology and Bionics", Vol. II, ed.
      Busnel, R.-G., Jouy-en-Josas, pp. 1052-1081.
Woodward, P.M., 1953, "Probability and Information Theory, with
      Applications to Radar", Pergamon Press, Oxford.

THE CONSTANT FREQUENCY COMPONENT OF THE BIOSONAR SIGNALS OF THE BAT, PTERONOTUS PARNELLII PARNELLII

O. W. Henson, Jr., M. M. Henson,

J. B. Kobler and G. D. Pollak

The auditory system of Pteronotus parnellii has been the subject of numerous studies but detailed knowledge of the emitted signals is limited. Previous studies utilizing tape recorded signals and sound spectrographs have shown that each emitted pulse consists of a brief upward sweeping initial FM component, a relatively long CF component and a brief downward sweeping terminal FM component. In addition there is a series of harmonics: the second harmonic (ca. 61 kHz) is the strongest. A study of the fine structure of this second harmonic CF component was undertaken to gain more critical information on pulse design, pulse emission patterns and the relationship of the signals to the sharply tuned cochlear microphonic (CM) audiogram.

CF component frequency measurements were made from filmed oscillographic displays of the output of a phase-locked-loop device which produced a dc voltage proportional to frequency (Smith et al., 1977) (Fig. 1). The CF component comprised from 70-83% of the total pulse. When the bats were not flying, the CF component usually reached a peak (maximum) frequency in the middle portion of each pulse and this frequency was maintained within 100 Hz for 33-65% of the total pulse duration. The beginning and terminal ends of the CF component represent transitions from the beginning and terminal FM components of the pulse and they can best be described as slow FM components, with frequency changes ca. 300-500 Hz; these slow FM changes constitute 15-40% of each pulse.

Records from flying bats were obtained by attaching the animals and a microphone to a pendulum with known velocity characteristics. Under these conditions the bats beat their wings, emitted pulses in characteristic way, and adjusted the peak frequency of the CF to compensate for Doppler shifts created by movement relative to a stationary target (Fig. 2). When a barrier was attached to the pendulum and positioned in front of the bats so that the echoes were not Doppler shifted, the bats did not adjust their pulse frequencies. Thus, Doppler compensation is entirely dependent on the perception of echoes.

Experiments with seven bats with chronically implanted electrodes showed that they consistently held the peak frequencies of the echoes at or below the tuned peak (best frequency) of the CM

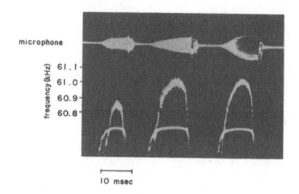

microphone

frequency(kHz)
61.1
61.0
60.9
60.8

|— 10 msec

Fig. 1. Oscillographic display of the pulses of <u>Pteronotus p. parnellii</u>. Upper trace shows the pulses as detected by a condenser microphone. The middle and lower traces show the output of a phase-locked-loop device where the CF pulse frequency has been converted into a dc voltage with gain settings of 100 Hz/div. and 1,000 Hz/div., respectively. Frequency calibration applies to middle trace. From Smith <u>et al</u>., 1977, J. Acoust. Soc. Amer. <u>61</u>, 1092.

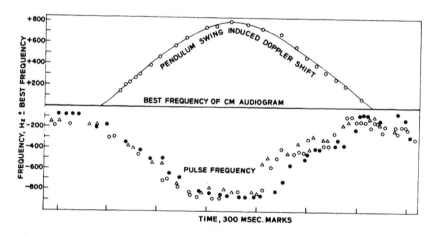

FREQUENCY, Hz ± BEST FREQUENCY

+800
+600
+400
+200
-200
-400
-600
-800

PENDULUM SWING INDUCED DOPPLER SHIFT

BEST FREQUENCY OF CM AUDIOGRAM

PULSE FREQUENCY

TIME, 300 MSEC. MARKS

Fig. 2. Graph of the changes in the peak frequencies of the emitted pulses in one bat before, during and after three forward swings (flights) on the pendulum device. Data for the different swings are indicated by the open and closed circles and triangles. The horizontal line marked "best frequency of the CM audiogram" corresponds to the tuned peak or threshold minimum and was determined for the same bat with the aid of chronically implanted electrodes. Note the close correspondence of the decrease in peak (pulse) frequencies with the increase in the peak frequencies of the echoes that should occur as a result of Doppler shifts imposed by the movements of the pendulum.

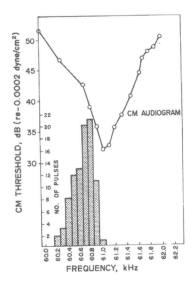

Fig. 3. Cochlear microphonic (CM)
audiogram for frequencies near
the 2nd harmonic of the CF component
of the emitted pulses and histogram
of pulse peak frequencies emitted by
the same bat when not flying.

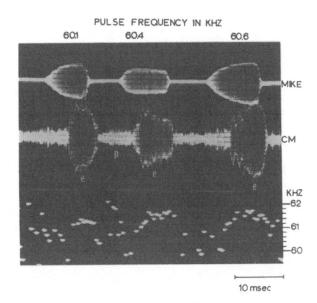

10 msec

Fig. 4. Cochlear microphonic responses recorded from flying
Pteronotus parnellii. Upper trace (MIKE) shows pulses detected
by a microphone; middle trace shows CM potentials to emitted
pulses (p) and Doppler shifted echoes (e); horizontal lines under
CM responses indicate the exact time of pulse emission relative
to CM responses. Dot display is output of period meter with
frequency scale indicated on right. Note the small responses to
emitted cries compared to echo-evoked potentials and the beats
created during periods of pulse-echo overlap.

audiogram;　in most cases the peak frequencies were 200-400 Hz below the tuned peak (Fig. 3).　In this frequency band the maximum amplitude of the CM potentials is higher and the intensity function curves are much more linear than at the best frequency of the CM audiogram.　These data explain how the ear of flying Pteronotus may show a relatively small response to the pulse CF component and an enormous response to Doppler shifted echoes which overlap the outgoing cry (Fig. 4).

(Supported by NIH Grants NS 12445 and NS 13276.)

THE AVOIDANCE OF STATIONARY AND MOVING OBSTACLES BY

LITTLE BROWN BATS, MYOTIS LUCIFUGUS

Philip H.-S. Jen, Yung H. Lee and R. Kelman Wieder

Echolocating bats sense their environments by emitting ultrasonic signals and listening to the echoes. The simplest way to
measure this ability is by the obstacle-avoidance test in which a
bat flies through an array of stationary wires and the numbers of
hits and misses are recorded. By extending this test and making
their bats fly between both stationary and moving obstacles, Jen
and McCarty (1978) reported that bets avoid moving objects more
successfully than stationary ones. Using the same experimental
method and more bats, we report here more quantitative data on the
sensitivity of the echolocation system of bats.

Two groups of little brown bats, Myotis lucifugus, caught in
early summer (20 bats) and early fall (9 bats) of 1978 were tested
in a room which is 4.6 m long, 2.2 m wide and 2.1 m high. An aluminum frame containing six vertically strung nylon monofilament lines
(1 mm in diameter) was placed in the center of the room. The distance between the lines was always 30 cm. The total opening of this
frame through which a bat flew was 1.4 m wide and 2.0 m high. By
means of an electric motor, the frame could be made to oscillate
with an amplitude of 14.5 cm in the horizontal direction parallel
to the alignment of the nylon lines. Any effect of the movement of
the frame itself on the bat's performance was eliminated by surrounding the frame with large wooden panels which were attached to the
walls, ceiling and floor. Thus a bat was confronted only with the
movement of the nylon lines, but not that of the frame. All surfaces of the wooden panels, ceiling and walls were covered with
convoluted polyurethane foam to reduce echoes. During the test, the
numbers of misses and hits were recorded when the obstacles were
stationary or oscillating at different rates. The average oscillating rate used for testing the bats caught in early summer ranged
from 0 to 28.7 cm/sec.

Assuming a bat could express its detection of moving obstacles
by increasing its score of misses, we have studied the sensitivity
of its echolocation system by statistically calculating the just-
noticeable moving speed of the obstacles which can be perceived by
the bat. That is, the minimum average moving speed of obstacles
which can be detected by a bat and behaviorally expressed by significantly increasing its score of misses relative to that obtained
from stationary obstacles. Nine bats caught in early fall were made
to fly through the array of obstacles which was oscillated at an
average speed of either 0, 2, 4, 6, 8, or 13.2 cm/sec.

The difference in scores of misses obtained under each test condition were then statistically compared (X^2 test, p = 0.05).

The results of these tests are summarized in Fig. 1. Bats can avoid moving obstacles with higher scores of misses than stationary ones. In both the fall and summer plots, all the data points obtained from moving obstacles are significantly higher than those obtained from stationary ones with only one exception. That is, in the fall plot, the data point obtained from obstacles moving at an average speed of 2 cm/sec is not significantly higher than that obtained from stationary ones. Interestingly, bats caught in early fall were able to avoid obstacles more successfully than bats caught in early summer.

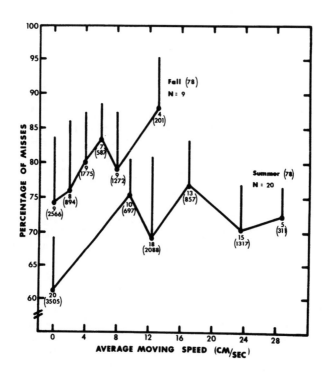

Fig. 1. The average percentage of misses obtained under different test conditions from two groups of <u>Myotis lucifugus</u> caught in early summer (20 bats) and early fall (9 bats) of 1978. Respectively, the ordinate and abscissa represent percentage of misses and average moving speed of the obstacles in cm/sec. Vertical bars at each point represent one standard deviation. Respectively, the numbers without and within parentheses under each point represent the total number of bats tested and the total number of flights under each test condition.

Three of the nine bats caught in early fall avoided moving obstacles more successfully than stationary ones, but the increases in their scores of misses were not significant. In the remaining six bats the scores of misses when presented with moving obstacles were significantly greater than when presented with stationary ones. The just-noticeable moving speed obtained from these six bats ranged from 2 to 6 cm/sec with an average of 3.7 ± 1.5 cm/sec.

That little brown bats can avoid moving obstacles more successfully than stationary ones may be associated with the fact that they rely upon moving prey for survival and they apparently concentrate more intently on moving objects while hunting. Since the flight speeds of insects preyed upon by bats are generally higher than 4.0 cm/sec (Johnson, 1969), bats apparently have no difficulty in detecting them. We suggest that a whole summer's practice of echolocation and the high activity of bats during the mating season may contribute to the superior performance of fall bats in avoiding obstacles.

Acknowledgements: Work supported by NSF and University of
    Missouri Research Council.

REFERENCES

Jen, P.H.-S., and McCarty, J.K., 1978, Nature, 275:743-744.

Johnson, C. G., 1969,"Migration and dispersal of insects by flight", Methuan, London.

ECHOLOCATION AND BEHAVIOR

Jean-Claude Lévy

An echolocating animal can adjust its signal in such a way as to receive a constant frequency for which the ear shows maximum sensitivity and ability to recognize echo characteristics. This is known as an "acoustic fovea". Is this adjustment the effect of automatic or purposeful behavior? Our assumption is that it is an automatic process involving not the echo but the original signal, which is automatically adjusted to obtain an optimal echo whose characteristics are therefore determined and known in advance.

Let $S(t)$ be the signal defined as a time function and $s(t)$ be the echo received after emission of the signal $S(t)$:

$$s(t) = O\{S(t)\}$$

where O is a transformation operator containing all target characteristics including range, size, shape, elasticity, and composition. To determine the operator O, both signal and echo must be known, but it is first necessary to determine which of the two is given and which is observed.

To illustrate the basic premises of the model, an analogy can perhaps be useful: we know the weight of an object that we lift from the strength of our muscles, which adjust automatically according to the value of this weight. In the same way, a bat could observe: "I am emitting loud signals, therefore my echo is weak" and not "my echo is weak, therefore I must emit loud signals". As frequency correction compensates for the Doppler effect and acts to cancel it, a bat could also observe: "my signal has a high frequency component, therefore my target is flying away from me" and not "my target is flying away from me, therefore I must emit a high frequency component".

Ranging would be an exception to this, as even if the optimal echo $s(t)$ is known, the time origin is not. For this reason, the signal alone is insufficient and the set, signal-echo, must be taken into account:

$$s(t + \tau) = O^*\{S(t)\}$$

where the unknowns are $S(t)$ and τ; $s(t)$ is given as the optimal echo which can be adjusted by the animal to obtain the most useful information about the target, e.g., frequency-time ambiguity.

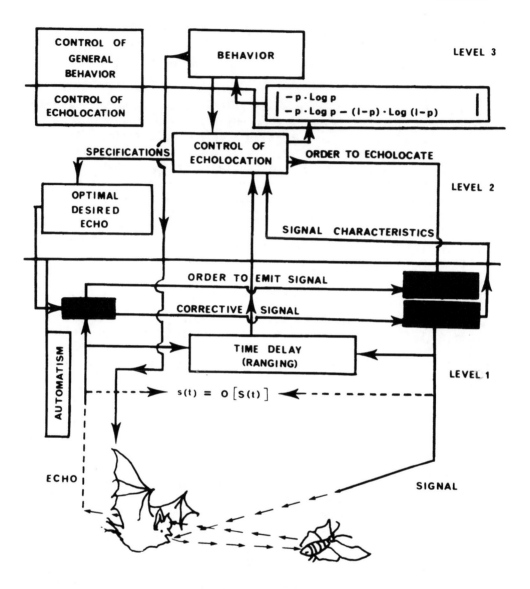

Fig. 1. Model of hypothetical neural integration.

Level 1: Automatism of the echolocation signal.
Level 2: Control of echolocation.
Level 3: Control of general behavior.

The principle of the operation is illustrated in Fig. 1 which presents different levels of hypothetical neural integration:

1. Automatism of the echolocation signal.
2. Control of echolocation.
3. Control of general behavior.

An observation made by Griffin (1958) lends support to the hypothesis of automatism and at the same time allows an additional element, the concept of probability relations, to be introduced between the second and third levels of the neural integration model. An _Eptesicus fuscus_, habituated to a room containing obstacles which it deftly avoided in its flights, began to collide with them when they were placed in different positions, in spite of the fact that it was emitting orientation pulses at the time. Apparently it was navigating from memory and did not bother to extract any acoustic information from its automatically emitted signals or their echoes.

This additional element can be described mathematically using the relationship given by Shannon's entropy. Let p be the a priori probability of finding an unobstructed passage in the room. The amount of attention that the bat pays to its echolocation pulses is proportional to the entropy term which, according to different hypotheses, can be given by one of the two formulae:

$$h = - p \cdot \log p$$

$$h^* = - p \cdot \log p - (1-p) \cdot \log (1-p)$$

If, on the basis of past experience, the bat is sure of finding the passage unobstructed, $p = 1$, h (and h^*) $= 0$, and the animal would pay no attention to its echolocation pulses which are automatically emitted. The phenomenon is not surprising as much of our own behavior is also based on probability relations and not on immediate perception of our surroundings. Consider, for example, the act of walking. When a particular route is familiar to us, we don't rivet our attention to the ground to make sure that it is always there when we take each step. Based on accumulated past experience, we are sure of finding the ground under our feet. But if one day a hole appears in the sidewalk, we will fall in it just as stupidly as the poor bat colliding with a newly moved and therefore unexpected obstacle in its familiar flight path.

THE FUNCTION OF LATERAL INHIBITION IN THE RECOGNITION OF

FREQUENCY SPECTRA

Jean-Claude Lévy

A biosonar system can recognize the nature of targets through an interpretation of echo characteristics. Humans are able to do the same thing only by striking an object and listening to the sound produced. The basic difference in the two procedures is that humans cannot do this at a distance.

An approximate estimation of frequencies is performed in the inner ear by the basilar membrane, whose movements are detected by the organ of Corti. The ear shows high precision in frequency discrimination and yet the response curve of the basilar membrane is very broad. We suggest that this precision is increased through the phenomenon of lateral inhibition.

The principle of lateral inhibition is well known in the study of the visual centers. Its function is to detect the boundaries of spots or illuminated fields by increasing the contast between bright and dark areas. It is also used in the auditory system but only for one dimension. Retinal cells which are fired by a field of brightness are replaced here by a segment of the organ of Corti. By increasing the contrasts and consequently the slopes of the response curves, lateral inhibition allows detection of boundaries with a precision greater than that required for a mere definition of the broad peak position.

Fig. 1 illustrates two frequency spectra, each representing 10 frequencies which are equidistant on a logarithmic scale. One spectrum is slightly higher than the other for upper frequencies, allowing theoretical differentiation of the two spectra. The global response of the organ of Corti to the two spectra (Fig. 2) is given by the summation of these curves, each with a weighting factor proportional to the log of the amplitude of the corresponding frequency. If $v(f,x)$ is the excitation of the organ of Corti for the frequency f, and $a(f)$ is the amplitude of that frequency, then the global response curve is:

$$r(x) = \Sigma a(f_i) \cdot v(f_x, x).$$

The organ of Corti cannot resolve the two spectra, and only one broad peak is observed.

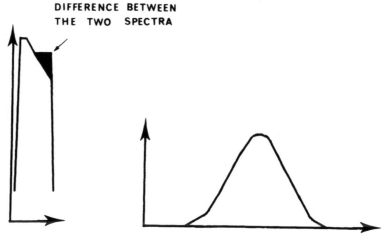

Fig. 1. Frequency spectra.   Fig. 2. Response of the organ of corti
                                         (no difference between the two
                                         spectra).

        The principle of lateral inhibition is illustrated schemat-
ically in Fig. 3. Columns of interconnected neurons whose relationships are excitatory extend from the organ of Corti. The neurons
of each line are connected by inhibitory relationships to the
neurons in adjacent columns of the following line. With such inhibitory effects integrated over broad areas of the organ of Corti,
stimulation of one area would directly inhibit activity in adjacent
less stimulated areas, thus leading to a sharpening and tuning of
the normally wide and flattened response curve of the basilar
membrane.

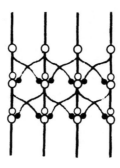

Fig. 3. Schematic representation of lateral inhibition.

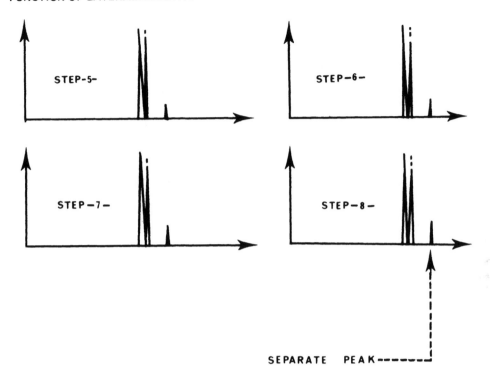

Fig. 4. The last four steps of lateral inhibition.

This procedure was simulated by means of an HF Mark 30 computer connected to a plotter (Fig. 4). The simplest possible curve
was taken for the characteristic response of each neuron: for a
positive input value, output equals input, while for a negative
input value, output is null. Following the action of lateral inhibition, the two spectra shown in Fig. 1 can be distinguished.
The higher spectrum is now characterized by a separate peak which
could theoretically be easily discriminated. No assumptions are
made about the meaning of these patterns. The only condition
stipulated is that the brain could easily distinguish and interpret
the two.

FURTHER STUDIES OF MASKING IN THE GREATER HORSESHOE BAT,

RHINOLOPHUS FERRUMEQUINUM

Glenis R. Long

Psychophysical studies have shown that the auditory system of man and other mammals seem to act like a series of bandpass filters which, with the exception of the lowest frequencies, increase linearly with log frequency. This author (Long, 1977) obtained an indirect measure of these bandwidths (critical ratios) using a classically conditioned response to shock and found that, while the critical ratios from Rhinolophus ferrumequinum were consistent with those from other mammals below 75 kHz, they decreased dramatically near the reference frequency. At that time it was not possible to specify the relation of these results to other modifications in these bats (see Neuweiler and Pollak, this volume). Further research was conducted using essentially the same method except that the noise was generated by multiplying 9.9 kHz low-pass noise by a pure tone giving rise to 19.8 kHz bandwidth 18 dB SPL/Hz noise centered at the frequency of the tone.

Fig. 1 shows that the tuning of the area of decreased threshhold associated with the reference frequency of these bats is not as pronounced in the masked thresholds as in the quiet but shows the same frequency dependency. The right-hand axis gives the estimated bandwidths based on Fletcher's formula (see Long, 1977). It should be noticed that the calculated bandwidths for frequencies around 81 kHz are greater than the 19.8 kHz noise available. Bruns (this volume and reviewed by Neuweiler, this volume) has developed a model of basilar membrane motion in this bat in which the relative motion of the basilar membrane is smaller at these frequencies due to a discontinuity in the nature of the membrane vibration. Neuweiler and Vater (reviewed by Neuweiler, this volume) found 8th nerve fibers which responded to frequencies above and below these frequencies but not to these frequencies. Möller (reviewed by Neuweiler, this volume) found that some collicular neurons show an absence of inhibition at these frequencies. The unusually large critical ratios at these frequencies may be related to these results and indicate that at least at the frequencies within the transition area of the basilar membrane this indirect measure is not a good measure at the frequency processing capabilities of the animal.

Direct measurements of the critical bandwidth were obtained using two different techniques: (a) Band Narrowing: pure tone thresholds were measured in noise of constant energy but decreasing bandwidth. The critical band was taken as the point where the threshold

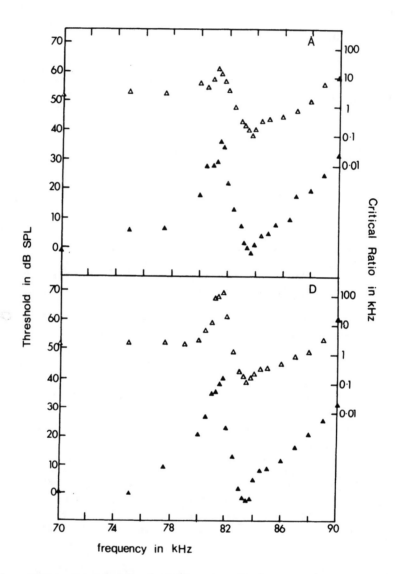

Fig. 1.   Quiet thresholds (solid triangles) and masked thresholds
(open triangles) as a function of frequency for two animals.   The
right-hand axis gives the calculated critical ratio bandwidths
from the masked thresholds.

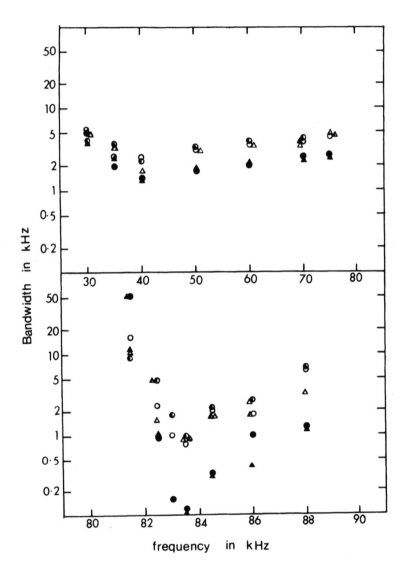

Fig. 2. Critical ratios (solid symbols) and two different measures of critical bands: band narrowing (open symbols) and two-tone masking of noise (half-filled symbols) as a function of frequency for two animals (D triangles, E circles).

stops increasing,   (b) <u>Two-tone Masking of Noise</u>; the threshold of a
100 Hz bandwidth of noise is measured as the separation of two pure
tones flanking it increases. The critical band is taken as the
point at which the thresholds begin to decrease as the tones move
outside the critical band. The results are presented in Fig. 2
along with the critical ratios from Fig. 1. Below 78 kHz the different measures of critical band give similar results that are approximately twice as large as the estimates of critical ratio and maintain a linear relation to log frequency except at low frequencies,
i.e., in this area this bat does not differ significantly from
other mammals, despite the increased innervation of the organ of
Corti from 70-80 kHz (see Neuweiler's review, this volume). Above
80 kHz the measurements no longer maintain a constant relation to
each other to the extent that near 81 kHz the critical ratios are
larger than the critical bands and at the reference frequency the
critical bands are six times the critical ratio. This may be due
to unusual patterns at vibration at the modified basilar membrane
and to unusual patterns of interaction between sounds as reviewed
by Neuweiler and Pollak (this volume). Additional data from this
author (in preparation) indicate that the bandwidths are asymmetrical
around the reference frequencies and approximately 2 kHz below it
(815 kHz).

    Schnitzler (this volume) suggests that these bats have fre-
quency resolution capacities of 30-60 Hz at the reference frequency.
The small critical ratios and critical bands at these frequencies
are consistent with these claims. As these results show this fine
frequency resolution capacity not only means that the bat is able
to detect small frequency changes but that he can also detect and
process these frequencies in high noise conditions. It also appears
that a Doppler shift compensating bat would not easily detect his
emitted sounds in a noisy environment.

Long, G. R., 1977, Masked auditory thresholds from the bat, <u>Rhino-
        lophus ferrumequinum</u>, J. Comp. Physiol., 116:247.

This work was conducted at the Polytechnic of Central London,
England. Now at Central Institute for the Deaf, St. Louis, Mo.,
U.S.A.

DOLPHIN AIR SAC MOTION MEASUREMENTS DURING VOCALIZATION BY TWO

NONINVASIVE ULTRASONIC METHODS

R. Stuart Mackay

Method one uses short pulses of 2 MHz sound in a sonar system. A probe on the subject's head projects impulses downward or inward. Echoes from successively deeper structures return to the probe to brighten a downward moving spot on a cathod ray tube. The spot jumps back to the top of the display and moves slightly to the right for the next pulse. Traced from left to right (Fig. 1) is the time pattern of movement of a line of internal reflecting structures, all correctly laid out in relative depth within the animal. Instead of being held stationary, the probe can be angulated on a small water-filled balloon (flexible coupling) to outline internal structures (form image) and thus achieve the desired aim for time recording.

The 6 Tursiops truncatus observed seemed undisturbed, though nonlinearities in their hearing mechanism in principle could allow them to sense the modulation envelope of this high frequency sound. Approximately 2000 distance measurements per second are performed (the pulse rate). Reflections from tissue-air interfaces are very strong but usually not directed back to the probe; what is seen is cessation of tissue echoes beyond the gas interface. Lower frequency sound gives deeper penetration and less resolution. Low intensity is harmless.

One can see which sacs move, when and how much, during various vocal activity. Fig. 1 shows motion of the anterior vestibular sac during a whistle. Distance from the probe on the skin surface to the intersection of sonic beam with sac is recorded. This diverticulum seems more than just a place to trap water leaks. Aiming in from down on the head allowed looking deeper under that sac at the nasofrontal sac.

There would be less possibility of animal disturbance if a steady input tone could give position information. It can. Movement velocity is recorded by measuring Doppler frequency shifts in the continuous signal returned from a steady outgoing 2 MHz. If desired, signal passage through an integrator records position (relative to initial position) as a function of time. In Fig. 2, special circuits recorded maximum sensed instantaneous advancing speed prior to and during clicks by an animal. Microphonics reduce if a bandpass filter follows the receiving transducer. All probes (Fig. 3) work in or out of water, and they can be fronted with a concave plastic lens to concentrate sensitivity on a smaller region. Fig. 4 suggests that signals from highly tilted surfaces might

933

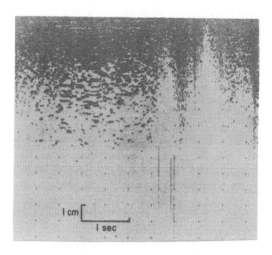

Fig 1 Echograph trace of vestibular
    sac motion.

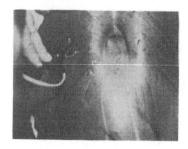

Fig 3 Probe on head of dolphin.

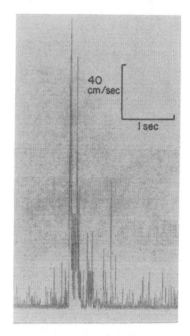

Fig 2 Doppler trace during
    clicks.

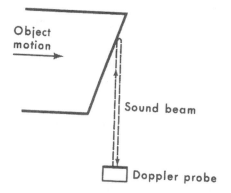

Fig 4 Exaggerated velocity is more
problem with pulsed than Doppler
units.

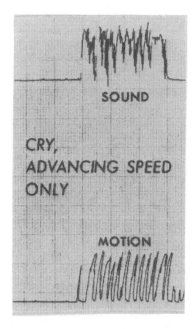

Fig 5 Similar patterns are
made with blowhole open or
closed.

exaggerate velocity if the geometry was unknown; in practice co-
herent echoes do not return in such a direction and one records component of motion along the beam, though in a pulsed system when
noting cessation of tissue echoes it can be misleading. Doppler
methods are especially useful for measuring the frequency of very
rapid motions since they can respond faster than periodically samp-
ling methods such as x-ray movies or pulsed ultrasound systems.
Thus they might also have application to dolphin hearing mechanism
studies where gas also seems involved.

Interpretation is aided by supplementary facts. Air interfaces in the vicinity of a primary sound source will be forced to
vibrate and modify the overall radiated pattern just as does a
piece of metal in the vicinity of a radio antenna (either trans-
mitting or receiving). In a human whistle air vibrates rather than
the lips and vortices are shed. I find it difficult to whistle at
pressures corresponding to below 25 m depth but it remains easy to
talk or to make "Bronx cheer" sounds. (A hyperbaric symphony orch-
estra has not been studied but mechanical whistles might be expected
to employ a different stimulation of resonances more related to
their click sequence generation.

Based upon the above, H. M. Liaw has begun his doctoral research with the methods. We find in Tursiops truncatus that before
or at the start of clicks the vestibular sac inflates on at least
one side and a nasal plug then vibrates. This sac acts as a reservoir with air coming and going but sound generates when air moves
up giving inflation. A "cry" such as in Fig. 5 when analyzed has
a low pulsatile frequency falling in the calculated resonance of
the vestibular sac which itself moves as the frequency is varied
and thus also probably acts as a resonator. Gross motion of the
other sacs has not yet been seen during clicks.

Some generalities on the nondestructive study of cetaceans,
including other details on ultrasound, have been given (Mackay,
1966).

Mackay, R. S., 1966, Telemetering physiological information from
        within cetaceans, and the applicability of ultrasound to
        understanding in vivo structure and performance, in: "Whales,
        Dolphins, and Porpoises", K. Norris, ed., U.C. Press,
        Berkeley.
Mackay, R. S., Rumage, W. T., and Becker, A., 1977, Sound velocity
        in spermaceti organ of a young sperm whale, in: "Proc. Second
        Conference on Biology of Marine Mammals", San Diego.

A THEORY OF THE SPERMACETI ORGAN IN SPERM WHALE SOUND PRODUCTION

R. Stuart Mackay

At a depth of 2500 m from where William Whitney recorded sperm whales sounding normal (unpublished), the density of air is about 1/4 that of water but the velocity of sound is approximately unchanged from the surface. (Air is not quite a perfect gas: at 250 atmospheres volume is 20% too large.) Acoustic impedance (velocity x density) thus is still less than that of water, and an air film remains a good sound reflector. Reflection of a sound pulse from the frontal air sac of a sperm whale showed excellent reflection with a phase change of 180° (inversion as in Fig. 1) suggesting the spermaceti organ might serve as a frequency controlling resonator or delay line analogous to an open ended rather than closed pipe. A reverberation model (Norris and Harvey, 1972) and a bugle model (Mackay, 1972) could be relevant. The energy release mechanism can involve air but frequency is depth independent if controlled by a liquid or solid resonator (or air if geometry could be fixed, viscosity changing little).

The exceptional anatomy of the sperm whale head was described by Raben and Gregory (1933) and by Norris and Harvey (1972). The cylindrical spermaceti organ has an air sac at each end, with a pair of lips in the forward one (Fig. 2-4), and is about 1/4 the length of the whale. The frontal sac of Physeter catodon provides a stable astigmatic concave-forward air film that focuses some sound in the vicinity of the Museau du Singe in the anterior sac, according to our measurements on one animal. The rear sac has a knobby back wall that can act as a spacer to stabilize the air film from mechanically distorting closed or dissolving even while pressurized to ambient by the lungs.

Vibrations of the lips could send sound down the organ where reflection would return the sound to shake the anterior sac and move air in synchronism through the lips, causing the lips to vibrate at the preferred frequency being generated just as do the lips of a trumpet player. In such modes of operation the lip end acts as a closed pipe with the sound period being 4 one-way transit times. The anterior sac would act as an air reservoir for recycling, as the analog of a trumpet mouthpiece, and as the coupling to the water and spermaceti. The organ is longer than wide and hence for the fundamental tone the proximal mirror is less than a wavelength across, making aim noncritical. The acoustic properties of the organ are rather like the surroundings and post mortem observations of channeling of sound at higher frequency can be misleading. However, if

937

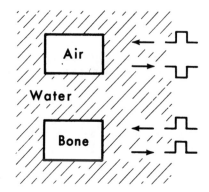

Fig 1   Fate of compression pulse after reflection
        from rather good reflectors, one acousti-
        cally hard and the other soft.

Fig 2   Spermaceti organ removed from over upper
        jaw has the Museau lips on front end and
        at the back had pressed against knobby
        tissue here held in place against front
        of skull by the hand.

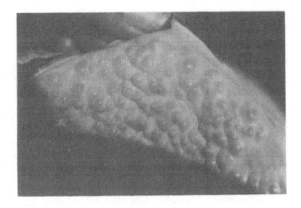

Fig 3   Covering of the back wall of the frontal
(posterior) sac.

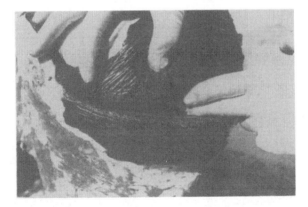

Fig 4   Museau du Singe lips after removal of
overlying front wall of anterior sac.

ducting of sound in the organ were involved, its conical taper would double the resonant frequency to one with a period of 2 transit times as if both ends were "open" or "soft", such as in a flute. (An oboe plays twice the frequency of the same length clarinet and overblows the octave rather than the twelfth.) Predicted frequencies are in the correct range, while variable lip tension, starting transients and short pulse length could cause the observed spread. Overtone emphasis depends on sizes and absorptions of all structures. Further observations will sort out alternatives, but the organ can control the frequency content of the clicks.

For the action to be like a trumpet blown for a few cycles, the natural frequency of the lips must be lower than the chamber resonant frequency, which seems true. If the column instead of interacting back on the lips merely emphasized certain frequency ranges (as with our use of vocal cords) then the start of each click would best communicate whale length. Another less likely model is that amplitude of a few cycles could be gradually built up in the column and then the stored cycles suddenly released as in a "Q switched" laser. Actual patterns seem not to favor models of passive axial reverberation of an initial click between two air sacs. While vortex shedding can take place in water (though not with water blown into air), I find it difficult to whistle at elevated pressure where "Bronx cheering" and speech with vocal cords remain easy; this may relate to why deep diving Physeter only clicks.

The velocity of sound in the spermaceti organ of a young animal at 35° C (a possible temperature) was V = 1.4 km/sec with a temperature coefficient from 20°-35° (while softening) of -5×10^{-3} km/sec/C°, versus the literature value of V = 2.6 km/sec. This velocity value relates a shorter whale to a given vocal frequency. If spermaceti density were very different from sea water, whales would have trim problems swimming level. Specific gravity was 0.86 here. Thus with the velocity of sound in tissue also being similar to that of water there is a fair impedance match for the release of sound, with sound being released in all directions from the forward sac (as is observed) and the monopole pattern being slightly modified toward a dipole pattern from the two sacs. The layered "junk" might help couple sound forward.

Norris, K.S., and Harvey, G. W., 1972, A theory for the function of
        the spermaceti organ of the sperm whale, in: "Animal Orient-
        ation and Navigation", Galler, Schmidt-Koenig, Jacobs and
        Belleville, ed., NASA SP-262.
Mackay, R. S., 1972, Discussion to above paper, 415-416.

HOW THE GREEN LACEWING AVOIDS BATS:

BEHAVIOR AND PHYSIOLOGY

Lee A. Miller

Bat sonar systems are primarily used for avoiding obstacles and detecting prey. But, not all prey are deaf to the ultrasonic cries of bats. Members of two orders of insects, namely the moths (Lepidoptera) and the green lacewings (Neuroptera), have evolved mechanisms for detecting and avoiding bats. In this poster I will present the results of studies designed to investigate the avoidance behavior of freely flying green lacewings (Chrysopa carnea) to hunting bats and the physiological mechanisms underlying this behavior. Detailed accounts are found in Miller and Olesen (1979) and Olesen and Miller (1979).

Interactions between hunting bats and freely flying green lacewings were studied in a large room using stop-motion photography and simultaneous recordings of emitted orientational cries. Pulsed ultrasound was also used. Electromyograms were taken from individual flight muscles while insects were in stationary flight.

Freely flying green lacewings often respond to hunting cries of bats by folding their wings and passively nosediving. Nosediving is an "early warning" response that takes the insect out of the bat's acoustic field. In the figure the insect begins its nosedive just prior to the bat's buzz (flash 5; the dots and numbers indicate stroboscopic flashes. L shows the bat's cries (repetition rate and relative cry duration) recorded by an ultrasonic detector to the left of the picture field). In this case the bat detected the insect too late to catch it. Bats caught less than 10% of the green lacewings they attacked and only 27% of the insects "buzzed" (see (1) in Table). Could green lacewings have other strategies for escaping bats after they had been detected? Should a falling green lacewing be detected, it can momentarily extend its wings and fold them again, continuing the fall. This "wing flip" constitutes a "last chance" response, which can be triggered by the high cry repetition rate in the buzz. The last chance response was studied in more detail using pulsed ultrasound. Green lacewings show other responses to ultrasonic stimuli such as changes in flight path and pattern. In addition the responses are variable, making them unpredictable. Hearing and responding to ultrasound gives intact green lacewings at least a 47% selective advantage over deaf green lacewings (see Table; destroying the ears does not interfere with flight). Green lacewings enjoy about the same selective advantage as tympanate moths with respect to bat predation (see Table) even though moth strategies differ from those of green lacewings.

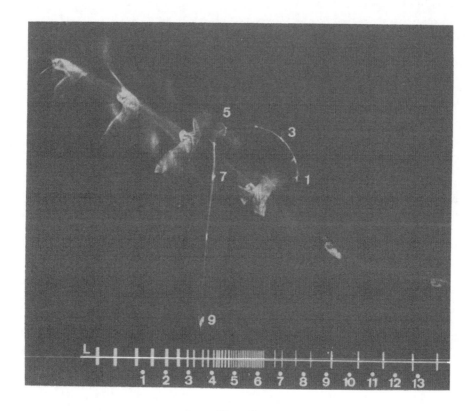

Fig. 1.

Table: Encounters between insectivorous bats and insects.

CATEGORY	(A) ATTACKS	(B) CATCHES	(C) MISSES	(D) BUZZES	(B/A) SELECTION PRESSURE	(B/D) CATCHING EFFICIENCY	SELECTIVE ADVANTAGE- REACTORS OVER NON-REACTORS
(1) REACTING GREEN LACE- WINGS	185[x]	18	167	67	0.10	0.27	$1-\frac{(2c/2a)}{(1c/1a)}$
(2) NON-REACTING GREEN LACE- WINGS	63[x]	33	30	37	0.52	0.89	0.47
(3) REACTING MOTHS	174	12	162	-	0.07	-	$1-\frac{(4c/4a)}{(3c/3a)}$
(4) NON-REACTING MOTHS	172	82	90	-	0.48	-	0.44

x : BASED ON THE NUMBER OF OCCURRENCES OF BUZZES AND HIGH CRY REPETITION RATES.

1 AND 2: DATA BASED ON ENCOUNTERS BETWEEN 3 BATS (TWO P. PIPISTRELLUS AND ONE M. BRANDTII) AND
             507 INSECTS (C. CARNEA). MAXIMUM ERROR IN THE DATA IS 10%. NON-REACTING GREEN LACEWINGS
             WERE EARLESS.

3 AND 4: DATA BASED ON FIELD STUDIES OF ENCOUNTERS BETWEEN 408 MEDIUM SIZED MOTHS AND UNIDENTIFIED
             BATS. COMPUTATION OF SELECTIVE ADVANTAGES IS FROM ROEDER & TREAT (1962), PROC. XI ENTOMOL.
             CONGR. 3: 7-11.

The activity patterns recorded from flight muscles correlated well with the behavioral responses from freely flying insects. Restraint has little influence on physiological responses. Green lacewings in stationary flight stop some or all of their flight muscles in response to pulsed ultrasound. If all muscles stop, the elevators stop last and start first upon the resumption of flight. This response would produce a nosedive with the wings in a dorsal position. The stopping of some muscles, but not others, would produce, irregular flight patterns. Physiological responses show variability both in which muscles stop and when they stop. Phase measurements between antagonistic muscles indicate that the stopping is produced by a temporary uncoupling of motor neurons from the continuously active flight pattern generator. The uncoupling mechanism is sufficient to explain the behavioral responses and variability in freely flying green lacewings. Variability, it seems, may increase the survival advantage of the prey by making its behavior more unpredictable for the predator.

I acknowledge the Danish Natural Sciences Research Council.

REFERENCES

Miller, L. A., and Olesen, J., 1979, Avoidance behavior in green
          lacewings. I. Behavior of free flying green lacewings to
          hunting bats and ultrasound, J. Comp. Physiol., 131:113.
Olesen, J., and Miller, L. A., 1979, Avoidance behavior in green
          lacewings. II. Flight muscle activity, J. Comp. Physiol.,
          131:121.

CYLINDER AND CUBE SHAPE DISCRIMINATION BY AN ECHOLOCATING BLINDFOLDED BOTTLENOSED DOLPHIN

Paul E. Nachtigall, A. Earl Murchison

and Whitlow W. L. Au

An Atlantic bottlenosed dolphin was trained to wear rubber eyecup blindfolds and to choose a cylinder as compared to a cube when the two targets were simultaneously presented. The animal was housed and tested within a circular fiberglass tank filled with filtered sea water. As may be seen in Fig. 1, the animal was trained to station within the tank and echolocate down a water-filled trough.

The targets were presented 2 m away from the animal, 40 cm below the surface, and 38 cm apart. The targets were placed between four 30 lb test monofilament lines attached, with stretchable elastic cord, to a wood frame.

On each of the 63 trials per session the animal was required to indicate the randomly predetermined right or left position of the cylinder by pressing one of two adjacent manipulanda. Each of the three different sized cylinders (with heights and diameters of

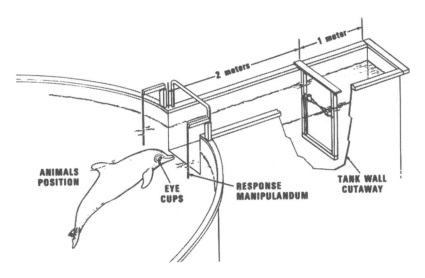

Fig. 1. Dolphin tank and apparatus.

4 x 4, 5 x 5, and 6 x 6 cm) was repeatedly paired with each of three different sized cubes (with heights, depths, and widths of 4 x 4 x 4, 5 x 5 x 5, and 6 x 6 x 6 cm). This variation in target sizes controlled for possible differences in overall reflectivity between cubes and cylinders. The order of presentation of the target pairs was randomly predetermined within each nine trials. That is, the 63 trials were broken down into seven nine-trial blocks. Each possible combination of targets was presented once within each nine-trial block.

Data in terms of the overall percent correct choice of the cylinder in each of the nine possible target combinations is presented in Fig. 2. The data indicate that the animal differentiated cylinders from cubes and that discriminability was influenced by target size. Comparisons of cubes and cylinders of the same size show increased discriminability of shape as target sizes increase.

Once the ability of the animal to discriminate cylinders from cubes was well-established, a probe technique was used to examine the effects of changing target aspect. Baseline performance trials were continued on 56 of the 63 trials per session but on the other seven trials one of the targets was either rotated or layed down horizontally. One probe trial was randomly presented within each

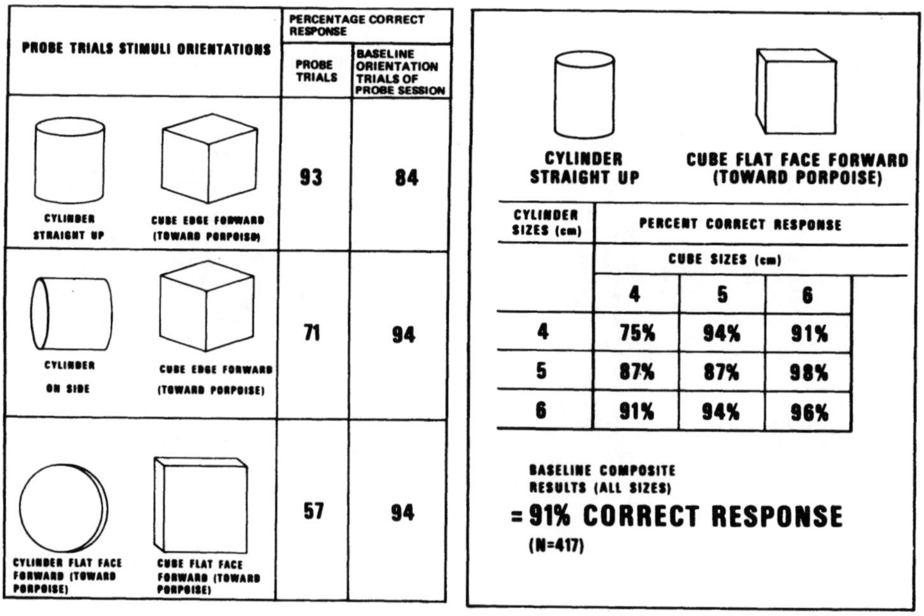

PROBE TRIALS STIMULI ORIENTATIONS	PERCENTAGE CORRECT RESPONSE: PROBE TRIALS	PERCENTAGE CORRECT RESPONSE: BASELINE ORIENTATION TRIALS OF PROBE SESSION
CYLINDER STRAIGHT UP / CUBE EDGE FORWARD (TOWARD PORPOISE)	93	84
CYLINDER ON SIDE / CUBE EDGE FORWARD (TOWARD PORPOISE)	71	94
CYLINDER FLAT FACE FORWARD (TOWARD PORPOISE) / CUBE FLAT FACE FORWARD (TOWARD PORPOISE)	57	94

CYLINDER STRAIGHT UP — CUBE FLAT FACE FORWARD (TOWARD PORPOISE)

CYLINDER SIZES (cm)	PERCENT CORRECT RESPONSE, CUBE SIZES (cm): 4	5	6
4	75%	94%	91%
5	87%	87%	98%
6	91%	94%	96%

BASELINE COMPOSITE RESULTS (ALL SIZES)
= 91% CORRECT RESPONSE
(N=417)

Fig. 2. Percent correct choice of cylinder.

nine trial block. A fish reinforcer was provided regardless of the animal's choice. As illustrated in Fig. 2, during the first two probe sessions the cylinder was placed straight up but the cube was rotated to present the edge forward. This change did not disrupt the animal's performance. During the next two sessions probe trials were presented with the cube edge forward and the cylinder layed down on its side. Performance declined to 71%. During the final two sessions both cubes and cylinders were presented on these probe trials with the faces forward. Performance declined to near chance level.

Following the completion of the experiment the targets were examined acoustically with a monostatic sonar measurement system. The echo returns for the various targets were processed, and resulting frequency spectra were obtained, by using a 1024 point fast Fourier transform. Comparisons of the target spectra failed to reveal consistent and obvious cues for the discrimination of shape.

Fifteen measures of target strength for each target, however, revealed an interesting possible cue for shape discrimination. The standard deviation of the target strength measures of the cubes straight up were 1.22, 1.01, and 1.01. The standard deviations of the cubes with the flat faces forward were much higher at 2.71, 5.92, and 6.18. Measures similar to the cubes were obtained from the cylinder presented with the flat face forward: 2.58, 4.93, and 5.86. These differences in variability paralleled the performance given by the animal. The animal most likely received repeated pulse echoes varying in amplitude when scanning across the flat surfaces of cubes or the tops of solid cylinders and relatively uniform pulse echoes when scanning across the curved portion of cylinders.

"RANGE" INFORMATION PROCESSING AT THE HIGHEST LEVELS OF THE AUDITORY SYSTEM.

William E. O'Neill and Nobuo Suga

Echolocating bats determine target range by measuring the time delay between the emitted biosonar pulse and the returning echo. We have found two types of neurons in a part of the auditory cortex of the mustached bat, Pteronotus parnellii rubiginosus, which demonstrate selective sensitivity to echo delays, and therefore, target range.

The biosonar signals consist of up to four harmonics, each of which contains a long constant-frequency (CF) component followed by a short frequency-modulated (FM) component. The duration and repetition rate of these signals become shorter and higher as the bat pursues and captures prey. In order to mimic as much as possible the actual biosonar signals and echoes for stimulation, we built two harmonic generators, which could produce up to three harmonics from a single fundamental frequency. We could, therefore, stimulate using some or all of these harmonics by switching them in or out at will.

Neurons in the FM processing area of the auditory cortex respond vigorously to a weak FM sound at frequencies near one of the non-fundamental harmonics of the biosonar signal, only if it is preceded by an FM sound mimicking the fundamental FM component of the emitted pulse in both amplitude and frequency. The responses remain essentially the same whether the pulse and echo contain all the harmonic components or just the essential FM components, or whether the CF components are produced continuously (no silent periods) or not at all (just FM). The time delay between the fundamental FM component (emitted sound) and the second, non-fundamental FM sound strongly influences the sensitivity of the neuron to the second FM sound (echo), and one can determine a best-delay for maximum excitation of a single neuron. Since in reality the delay is related to the distance to the target, these neurons are range-sensitive.

When presented with pulse-echo paired stimuli, tracking neurons respond to echo FM's from a wide range of delays (target distances) at repetition rates mimicking the search and approach phases of target-oriented flight, but the best delay decreases dramatically during the terminal phase, just prior to target interception. Correspondingly, their sensitivity to targets at other distances decreases sharply. Thus they appear to follow, or track, the target as the bat approaches it.

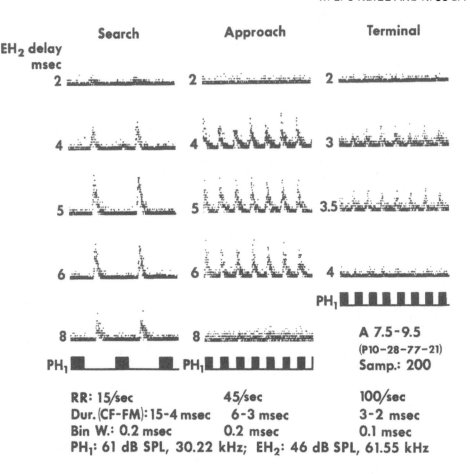

PST histograms illustrating a tracking neuron are shown in the accompanying Figure. Note that as the repetition rate (RR) of the stimulus pairs increases, both the range of echo delays over which the neuron responds (EH_2 delay) and the best delay decreases. No responses were elicited to pulses or echoes delivered alone, or to pulse-echo pairs with delays shorter than two msec.

Range-tuned neurons, on the other hand, do not change their best delays with changes in repetition rate and duration of the stimuli. They respond to targets located only at certain distances, regardless of the stage of echolocation. Although best delay remains similar in any one cortical column, it differs among columns in the FM processing area. We have found that the best delay increases systematically in recordings from oblique electrode tracks moving rostro-caudally through the FM processing area. Range-tuned neurons are therefore arranged on an "odotopic" axis, in which target range is systematically represented by place on the cortical surface. The average change in best delay from neuron to neuron along this target range axis is 0.116 msec, corresponding to an average resolution of 2.0 cm. This correlates well to behavioral evidence for the acuity of target range discrimination in bats. The bats Eptesicus fuscus, Phyllostomus hastatus, Pteronotus suapurensis, and R. ferrumequinum are able to discriminate 1.2 to 2.5 cm range differences at an absolute distance of 30 and/ or 60 cm (Simons, 1971; Simmons, Howell and Suga, 1975). This would also correspond to our finding, if we assume that the rate of change in best range, 2.0 cm per neuron, is the theoretical limit of just noticeable difference in distance.

It is puzzling that the fundamental (H_1, in particular FM_1) of the orientation sound is so critical to the response of range-tuned neurons in spite of the fact that H_1 is always much weaker than the second and third harmonics and is sometimes barely detectable in recordings in our laboratory. We propose that this is an adaptation to reduce the probability of jamming by the sounds of echolocating conspecifics. When we delivered pairs of orientation sounds and echoes without H_1, we found no range-sensitive neurons in the FM processing area. This means that range-sensitive neurons are probably not excited by combinations of orientation sounds and/or echoes produced by conspecifics flying nearby. To excite range-sensitive neurons, H_1 must stimulate the ears prior to an echo, in spite of its weakness in the emitted sound. This suggests that H_1 produced by the vocal cords stimulates the animal's own ears by bone conduction, while it is not emitted at a significant amplitude possibly because of suppression by vocal-tract anti-resonance. In nature, range-sensitive neurons would be selectively excited only when the animal itself emits orientation sounds and echoes return with relatively short delays. Without this property, target ranging would be frequently impaired by conspecifics. Thus these neurons are jamming-tolerant in most situations.

Supported by NSF grant BMS 75-17077 and BNS 78-12987 to Nobuo Suga, and NINCDS (PHS) training grant 1-T32-N507057-01 to William E. O'Neill.

References

Simmons, J.A. (1971). The sonar receiver of the bat. Ann. N.Y. Acad. Sci. 188, 161-174.
Simmons, J.A., Howell, D.J., and Suga, N. (1975). Information content of bat sonar echoes. Am. Sci. 63, 204-215.

THE FUNCTIONAL ORGANISATION OF THE AUDITORY CORTEX IN THE

CF - FM BAT RHINOLOPHUS FERRUMEQUINUM

Joachim Ostwald

Greater Horseshoe Bats, Rhinolophus ferrumequinum, use orientation sounds consisting of a long constant frequency (cf) part, followed by a short, downward frequency modulated (fm) part. By lowering their emission frequency during flight they compensate Dopplershifts and keep the cf-part of their echoes in a very small frequency band that is characteristic for each individual animal. Thereby they create a constant carrier frequency on which the fast, complex amplitude and frequency modulations are superimposed that are produced by the movements of targets, for instance the wingbeat of flying insects. Those modulations are characteristic for each insect species and size and may therefore be used to classify or even identify prey and to distinguish them from background clutter.

Multi-unit recordings show that the auditory cortex of Rhinolophus is especially adapted to the processing of echoes. Units with best frequencies (BF's) in the frequency band of the orientation sounds are strongly overrepresented in the tonotopical organisation (Fig. 1). There are two large areas that are concerned with the processing of the cf- and the fm-part of the Dopplershifted echoes respectively (shaded in Fig. 1). In both areas the frequency axis is stretched compared with other areas. In the fm-processing-area and in areas not concerned with echo processing neurons with equal BF's are organised in dorsoventral slabs with raising frequency from posterior to anterior. In the cf-processing-area frequencies between resting frequency and 2 kHz above are represented which are the frequencies occuring mostly in the Doppler-shifted cf-parts of the echoes. Here equal-BF-contour-lines are bent around each other in a semicircle arrangement with the resting frequency at the outside. When frequencies are normalized to the individual resting frequency of the animal, this pattern stays constant even for large variations in resting frequency. A similar disproportionate representation of echo frequencies was found in the auditory cortex of Pteronotus parnellii (summarized in: Suga & O'Neill, review at this conference). In contrast to Rhinolophus equal-BF-contour-lines in the Doppler-shifted-CF processing area of Pteronotus are organised in closed circles and with a reversed frequency arrangement.

The tuning curves of neurons with BF's in a frequency band of $\pm$ 2 kHz around resting frequency are extremely sharp (Q 10-dB values up to 400). They have high frequency selectivity and therefore are well suited for coding even small frequency modulations. With de-

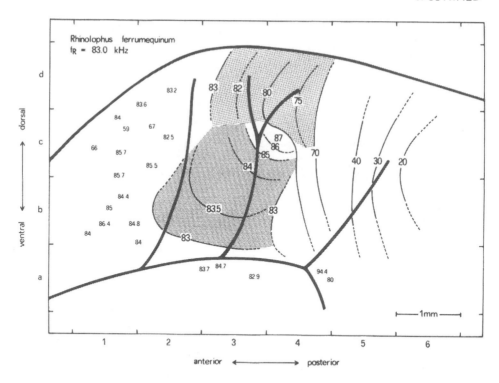

Fig. 1. Tonotopical organisation of multi-unit best frequencies recorded in 5 bats. Measured frequencies have been normalized to the resting frequency of the individual animal. For plotting the resting frequency was set to a mean value of 83.0 kHz. The shaded areas represent the cf- and the fm-processing-area respectively.

creasing BF the Q 10-dB value becomes smaller. For BF's in the frequency band below 70 kHz, that is not used for echolocation, Q 10-dB values are smaller than 20 which is the maximum value reached in other mammals. Sharp tuning of neurons with BF's around resting frequency is found in all centres of the auditory pathway of Rhinolophus studied so far and is probably due to a sharp mechanical filter in the cochlea (summarized in: Neuweiler, review at this conference).

The response pattern of these neurons changes when sinussoidally frequency modulated stimuli (SFM) down to a frequency shift of 0.05 % are used. With larger modulation depth phase locked responses occur. The degree of phase locking depends not only on the modulation depth but also on the position of the carrier frequency within the pure tone tuning curve. The response to complex frequency and amplitude modulations caused by the wingbeat of flying moth is phase locked to the wingbeat and differs from the response to SFM (Fig. 2 B, C). Both responses can occur at carrier frequencies where there is no response to the unmodulated carrier itself (Fig. 2 A - C).

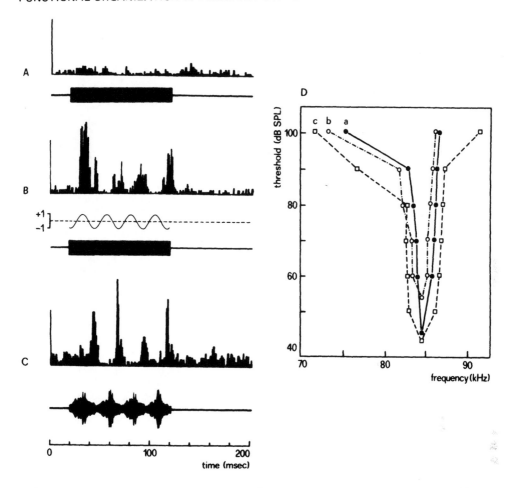

Fig. 2. Response pattern of cortical neurons to a pure tone, 82.0 kHz, 80 dB SPL (A), to a sinussoidally frequency modulated stimulus (SFM), modulation depth ± 1 kHz (B), and to the reproduced echo of a flying moth (C) on the same carrier frequency. In the complex moth echo frequency and amplitude modulations are present. The response is phase locked to the wingbeat. (D) Threshold curves of the same unit with (a) pure tone, (c) SFM, and (b) moth echo stimulation. Plotted are frequency and intensity of the carrier.

Compared to the pure tone tuning curve (Fig. 2 D.a) the threshold curve for SFM stimuli, modulation depth ± 1 kHz, (Fig. 2 D.c) is 1 kHz wider on both the high and low frequency side. For moth echo modulations there seems to be a special mechanism of processing in the unit shown in Fig. 2. Although in this stimulus there is an asymetrical widening of the frequency band present with each wing-beat of -2 to +1 kHz (see Schnitzler & Henson, review at this conference, Fig. 16), carrier frequencies below the pure tone BF are more effective in eliciting a response (Fig. 2 D.b).

TURSIOPS BIOSONAR DETECTION IN NOISE

Ralph H. Penner and James Kadane

Biosonar detection experiments conducted in the ocean are confounded as the range from the animal to the target is increased because added interference, such as fish schools, occlude the "field of view". This experiment was designed to test deduction performance as a function of noise at a fixed range of 16.5 meters, thus reducing the range variability problem. Echolocation data were collected from two Tursiops truncatus, named Ehiku and Heptuna, in Kaneohe Bay, Hawaii, during November 1978. They were conditioned to respond discretely in a two alternative forced choice detection problem of 0.5 a priori random presence or absence of a 7.65 cm steel water-filled sphere (Ts. -32 dB). The sphere, displayed one meter below the surface and 16.5 meters from the porpoise at. scan station, assured stable low effort baseline detections in the absence of added noise. Detection accuracy, pulse train duration,, and response latency data were recorded in response-terminated trials. Five levels of white noise, 67, 72, 77, 82, 87 dB rel μPa in a 1 Hz band, were presented in 10 trial blocks, each level occurring twice in each 100 trial session. There were a minimum of 220 trials at each sound level. Ambient noise, which is extremely high in Kaneohe Bay, was about the same as the lowest white noise level used in the experiment.

A white noise source was located 4 meters from the porpoise and directly in line with the sphere or the position occupied by the sphere. It was activated before an acoustically opaque screen was lowered which permitted an acoustic view of the stimulus area, and deactivated after a response on one of the manipulanda.

Although there was no significant change in response accuracy over baseline with the 67 dB noise, response bias became neutral (Fig. 1) and there was an abrupt increase in the number of clicks per click train (from 25 to 50) emitted by the two animals (Fig. 2). Overall performance degraded with both animals as the noise level increased. Ehiku's detection/rejection ratio remained neutral, whereas Heptuna showed a liberal bias in his decision process as the noise level increased. It was interesting to find that the click count ("echolocation effort") (Fig. 2 and 3) and response latency (Fig. 4) both increased with both animals until the noise exceeded 77 dB. At 82 and 87 dB, where the overall performance was near chance, the click trains got shorter and latencies longer.

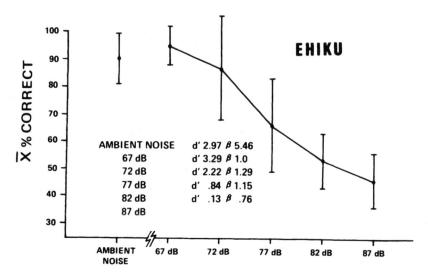

Fig. 1. Response Accuracy Ogive and Signal Detection Parameters,
         Sensitivity d[1] and Response Bias β.

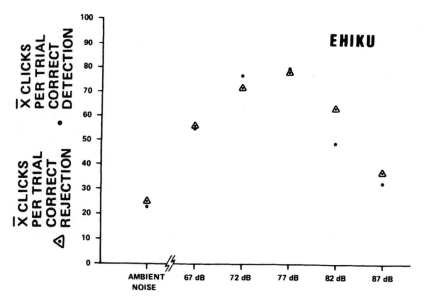

Fig. 2. $\overline{X}$ Clicks per Trial

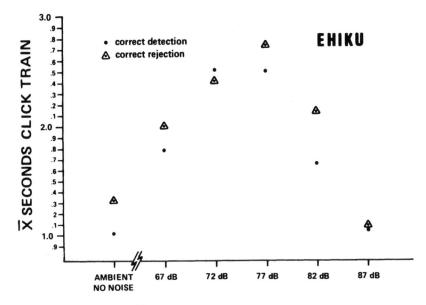

Fig. 3. X̄ Seconds Click Train per Trial

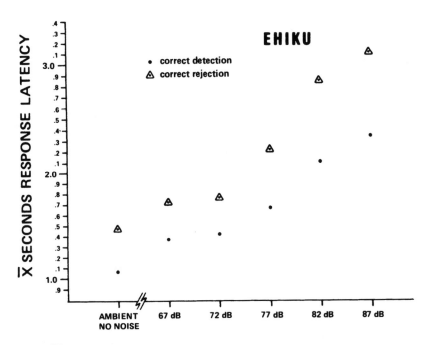

Fig. 4. X̄ Seconds Response Latency, Time from End of Click Train
to Activation of Response Manipulanda.

CORRELATION ANALYSIS OF ECHOLOCATION PULSES

J. David Pye

Following the lead of D. A. Cahlander (1967) and more recently of K. J. Beuter (this conference), radar ambiguity diagrams of bat pulses must now be regarded as an important extension to merely descriptive or acoustic analyses. The sonar properties of such signals are not intuitively obvious for several reasons:

(1) Their frequency and amplitude patterns differ from those of 'worked examples' given in radar texts and, indeed, often defy the engineers' design constraints (eg. envelope shaping, unusual f.m. patterns and the presence of harmonics); such signals may be difficult to define for purely mathematical analysis. (2) Several of the assumptions of radar theory (or even under-water sonar theory) are not valid for ultrasound in air: the fact that band-width is large compared with the centre frequency means that Doppler shift must be treated as a _factor_ (Cahlander) and not approximated by an incremental offset, so that for example, Doppler tolerance is achieved by linear _period_ sweeps (Altes and Titlebaum, 1970), not by linear f.m. (3) Propagation attenuation in air is strongly dispersive, the atmosphere acting as a low-pass filter (Griffin, 1971), and this must affect many long-range 'cruising' pulses (Fig. 1). (4) The back-scattering cross-sectional area of many targets is dispersive (eg. small insects are high-pass Rayleigh scatterers, Fig. 2). At certain ranges, depending on frequency and water vapour concentration, these two effects cancel out to give a 'flat' echo spectrum (Fig. 3). Bats presumably allow for all these effects in the evolution of their signals and in their methods of signal processing.

Our hybrid, true-Doppler correlator, described by J.A.T. Halls (poster, this conference) is therefore being used to examine:

(1) Pulse structures from our own recordings of 118 species of echolocating birds and bats. (2) The changes of pulse structure shown by many species under different conditions. (3) Artificial signals which simulate real pulses (Pye, 1967), so that noise is minimised and each of the pulse parameters can be varied independently. (4) Signals propagated over different distances. (5) Real echoes reflected from various targets. Correlation also offers the possibility of locating sources from short baselines and so of tracking individual bats in flight. Clearly such methods have great potential for the study of echolocation mechanisms in both theory and practice.

References: please see author's Review chapter.

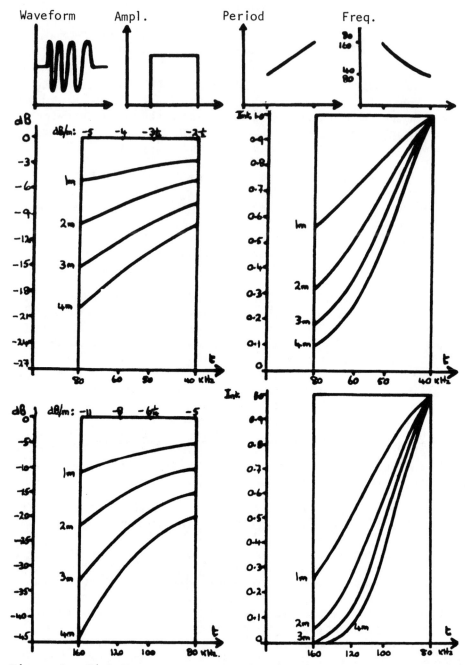

Figure 1. High frequency attenuation in air. Top; two model pulses
with square envelopes and linear period modulation over one octave.
Middle: the effects of propagation over 0-4 m in air with 1% H$_2$O,
at 80-40 kHz (dB left, normalised intensity right). Bottom: the
same at 160-80 kHz. Higher water vapour content gives greatly
increased effects. Data from Griffin (1971).

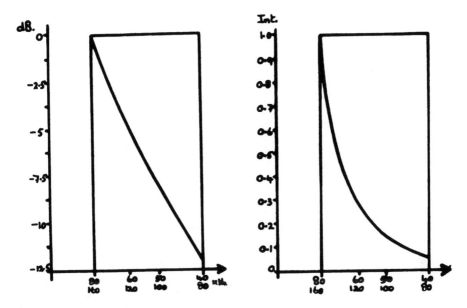

Figure 2. Rayleigh back-scatter from a small sphere ($\lambda \gg \pi d$) for the model pulses of Fig. 1.

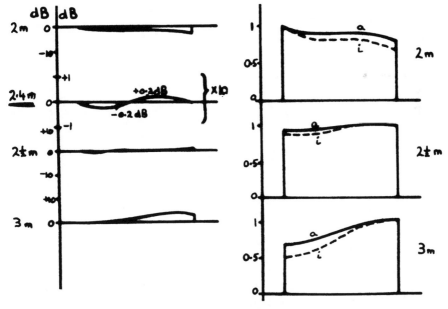

Figure 3. Low-pass attenuation and high-pass Rayleigh scattering can almost cancel out at certain ranges to give a 'flat' echo spectrum. For 80-40 kHz in 1% H_2O this occurs at 2.4 m; left: the echo in dB (at 2.4 m expanded x 10 to show ± 0.2 dB); right: intensity. Similar 'flat' returns occur at 0.9 m for 160-80 kHz in 1% H_2O and at 1.5 m for 80-40 kHz and 0.6 m for 160-80 kHz in 3% H_2O.

THE COCHLEA IN PTERONOTUS PARNELLII

Ade Pye

Pteronotus parnellii provides an exceptional example among the New World bats so far studied in being able to compensate for Doppler shifts (Schnitzler 1970) and in possessing a cochlea which is sharply tuned to just above emitted frequency at rest (Pollak et al 1972). Structurally the cochlea exhibits a number of specialized features (Pye,A. 1967 and 1978). For example, the basal turn is much enlarged, including an enormous second half-turn which exhibits the largest scala vestibuli found in any of the bats studied. It has recently been reported by Pye,A. (F.I.B.R.C., Albuquerque 1978) that another special feature of the basal turn is the presence of an un-identified substance in the scala tympani, see Fig.1. It is located mainly at the medial and posterior walls of this scala and has also now been found to reach around the round window and the cochlear aqueduct. A smaller area is situated anteriorly, see Fig. 2. This substance has been found in every cochlea of Pteronotus parnellii studied (24 in all) from various locations in Trinidad and Panama. But it is completely absent in Pteronotus davyi and all other bats examined to date (68 species from 14 families).

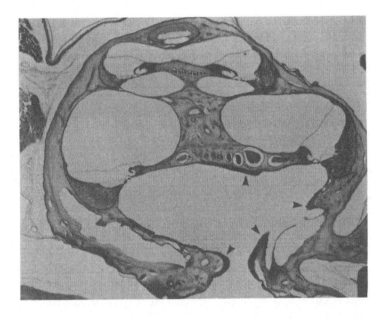

Fig.1. Horizontal section through the cochlea of Pteronotus parnellii (x 29). The arrows indicate the position of the substance. Note lysis of the cartilage situated in the bony cochlear wall on the left side.

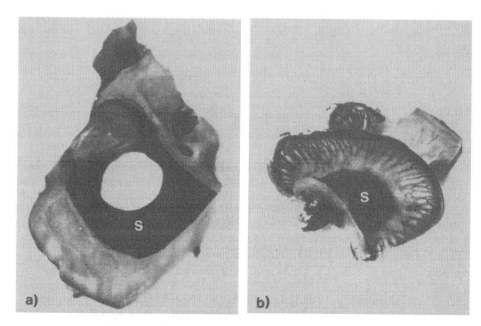

Fig.2. Micro-dissected cochlea of <u>Pteronotus parnellii</u> (x 16)
       S indicates the substance. a) shows the posterior part of
       the substance around the round window and onto the medial
       walls, b) shows the anterior part of the substance. Notice
       the enormous basal turn when compared with the second turn.

Under light microscopy it appears amorphous in structure, measuring
between 30µm and 40µm in thickness and being surrounded by a
distinct membrane in places. Further histological staining has
shown it to stain a) negative with Gram, b) positive with Amido
Black (specific for haemoglobin), c) red with Masson's Trichrome
(red for blood corpuscles and muscles), d) with Congo Red and
Van Gieson's (Elastin and Haemotoxylin) the same as for red blood
corpuscles in the surrounding vessels. Electron microscopic studies
also indicated it to be amorphous, but now new observations have
shown that there are distorted red blood corpuscles on the surface,
with massed haemoglobin underneath, see Fig.3. This is a surprising
finding and yet it agrees with the results obtained from histological
staining viewed under light microscopy. The massed haemoglobin
is obviously not randomly distributed, since it appears in precise
locations only. But as lysis of the cartilage situated in the bony
cochlear wall (see Fig.1.) also seems at present to be confined
to this species, the two phenomena could possibly be related.
Since this symposium only one foetus and one baby bat of this
species were available for histological study and neither showed
the presence of the massed haemoglobin nor lysis of the cartilage.

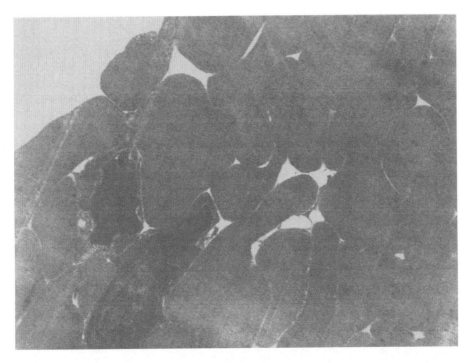

Fig.4. Electron micrograph of the substance, showing distorted red
       blood corpuscles on the left, with massed haemoglobin
       underneath on the right. (x 9250) (Courtesy of Dr.A.Forge)

It is obviously of great importance to trace the evolution of this
massed haemoglobin through foetal and young development to the
adult cochlea. Further investigations are being carried out, but
at present the definite origin and possible functional significance
of the massed haemoglobin remains unknown.

Pye, A. 1967. The structure of the cochlea in Chiroptera. III
       Microchiroptera: Phyllostomatoidea. J. Morph. 121, 241-254

Pye, A. 1978. Aspects of cochlea structure and function in bats.
       Proceedings of the Fourth International Bat Research
       Conference, Kenya National Academy for Advancement of
       Arts and Sciences, Kenya Literature Bureau.

ACTIVITY OF THE RECURRENT LARYNGEAL NERVE DUE TO THE PRODUCTION

OF ULTRASONIC ECHOLOCATION SOUNDS IN THE CF-FM BAT,

RHINOLOPHUS FERRUMEQUINUM

Rudolf Rübsamen

The Greater Horseshoe bat, Rhinolophus ferrumequinum emits a stereotyped echolocation sound consisting of a long constant-frequency portion (CF) terminated by a short frequency-modulated part (FM). During echolocation the frequency of the orientation pulse is controlled with high accuracy (Dopplershift compensation) and the duration of the call is shortened while the bat is approaching its targets.

The orientation pulses are produced in the larynx, which is innervated by two branches of the vagal nerve, the superior laryngeal nerve and the inferior laryngeal nerve. The superior laryngeal nerve, innervating exclusively the cricothyroid muscle, controls the frequency of the orientation sound. The inferior laryngeal nerve innervating the rest of the intrinsic laryngeal muscles, seems to be responsible for the temporal control of sound emission.

Simultanously recorded respiration signals and nerve activity of the recurrent laryngeal nerve show that this nerve is active during respiration as well as during vocalization (Fig. 1 and 2). The respiration is highly adapted to the demand of vocalization. Even in the quiet respiration without vocalization the expiration phase is a short pulse of air lasting about the length of an orientation call (50 msec). The following inspiration phase is longer and lasts about 150 msec. In the experimental situation, where vocalization was elicited by electrical stimulation, the vocalization does not change the rhythm of respiration and is strictly correlated with the expiration phase.

During vocalization a stereotyped activity pattern superimposes the respiratory activity (Fig. 3 and 4). 20 msec prior to the onset of vocalization the activity of the inferior laryngeal nerve reaches its first maximum. This level of activity is roughly maintained until 10 msec before the sound terminates. A short notch of reduced activity is then followed by a second strong activity maximum that parallels the termination of vocalization. The subglottic pressure is build up shortly before the onset of vocalization, it is maintained during the emission of the sound and drops at the end of the outgoing vocalization.

Viewing subgottical pressure and inferior laryngeal nerve activity
together suggests that the activity of the inferior laryngeal nerve
might be responsible for the closure of the larynx to build up
subglottic pressure prior to the vocalization and for the opening
of the larynx to reach the respiratory state at the end of the sound.

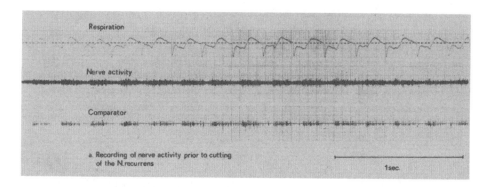

Fig. 1   Registration of respiration activity of the recurrent
          laryngeal nerve.

          In the upper trace respiration is recorded. The middle
          trace represents the registration of the nerve activity
          and the lower trace shows the comparator signal of the
          nerve registration.

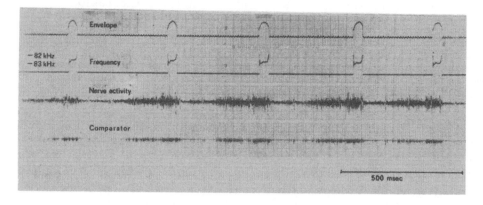

Fig. 2   Recording of vocalization activity in the recurrent
          laryngeal nerve.

          Continuous recordings of the envelopes of the orientation
          calls, their CF-frequency, the gross activity of the nerve
          and the comparator signal for histogram-formation are
          shown.

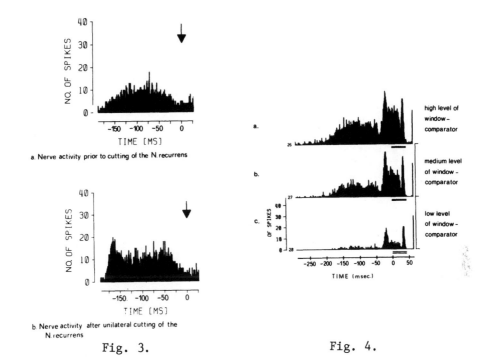

a. Nerve activity prior to cutting of the N. recurrens

b. Nerve activity after unilateral cutting of the N. recurrens

Fig. 3.   Fig. 4.

Fig. 3   Ante-stimulus-time-histogram of nerve activity of the
         recurrent laryngeal nerve during respiration without
         concurrent vocalization.

         a. Nerve activity before the transection of the recurrent
         laryngeal nerve.
         b. Nerve activity after unilateral transection of the
         recurrent laryngeal nerve.
         The start of inspiratory phase is marked by an arrow.
         Each histogram is composed of 40 sweeps.

Fig. 4   Ante-stimulus-time-histogram of nerve activity of the
         recurrent laryngeal nerve during vocalization. The
         histograms are triggered by the onset of vocalization. The
         position of the orientation calls is represented by
         double-bars. All three histograms are summed up from the
         same sample of orientation calls using different discri-
         minator levels (a,b,c). The histograms show that indepen-
         dent of the discriminator level the same discharge-pattern
         of the recurrent laryngeal nerve during vocalization is
         obtained.

SINGLE BRAIN STEM UNIT RESPONSES TO BINAURAL STIMULI SIMULATING MOVING SOUNDS IN RHINOLOPHUS FERRUMEQUINUM

Peter A. Schlegel

Echolocating bats are essentially dependent on the exact determination of target locations and their changes relative to the bat. Echoes reflected from any space location are determined by the distance and the vertical and horizontal (azimuth) angle to the bat's longitudinal axis.

The generally accepted theory of directional hearing assumes neural mechanisms that integrate inputs from both ears (binaural hearing). The most significant stimulus cue concerning the azimuth angle of a sound source, i.e., echoes returning from targets, for bats is the intensity difference between the two ears.

Directionality of Rhinolophus' hearing system

From neurophysiological data on the directionality of the ears (Schlegel, 1977a) one can derive "interaural pressure difference" curves (IPD) coding azimuth angles between ±40° from straight ahead direction, probably the biologically most significant range. This convertability offers the advantage of replacing free field stimulation via loudspeakers by binaural intensity combinations applied through earphones (Schlegel, 1977b).

Neurophysiological properties of binaural neurons

Single unit recordings from binaural brain stem levels revealed that in Rhinolophus, similar to other mammals studied, the most prominent binaural mechanism consists in a subtraction of the neural inputs between right and left ears: stimuli delivered to one channel provide excitation, whereas stimuli into the other inhibit excitation so that ideally the resulting activity is a directly proportional measure of the azimuth angle. This general relationship holds for the whole excitatory/inhibitory dynamic range of a unit but is quantitatively not independent of absolute intensity levels; yet one can state that brain stem units are suited to even precisely code angular positions (best incremental limen estimated as to ± 1°: Schlegel, 1977a).

Coding of moving sounds

Since bats hunt while flying one has to consider relative movements between bat and target as biologically highly relevant features

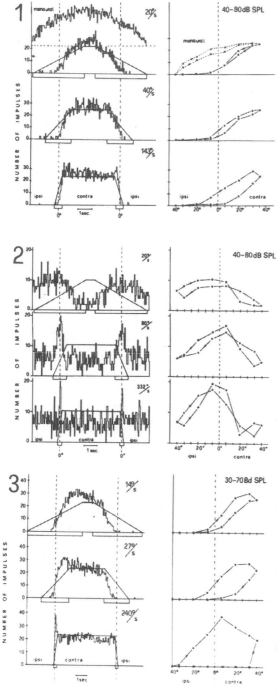

fig. 1-3

A. Time representation

Time course (abscissae, 1s
time bar; 2oo bins, binwidth
46.7 ms, 5 runs averaged) of
single unit activity (ordina‑
tes, number of impulses/bin)
when trapezoidal intensity
changes, indicated by thin
solid lines, are applied simu‑
lating rotating sound sources.
Histograms gained with diffe‑
rent sweep speeds are display‑
ed vertically (up to down, in‑
creasing velocity as marked);
absolute intensity used as in‑
dicated. The histograms are
aligned one above the other
to the reference of o^{o} appa‑
rent azimuth angle (dashed
vertical lines). Curves were
measured binaurally unless
otherwise labeled. Wordings
"contra" and "ipsi" stand for
virtual sound source locations
refered to the recording side.

B. Angle representation

Same data replotted in angular
representation (abscissae, azi‑
muth angle, ^{o}; ordinates as in
A) for the two opposite move‑
ment directions as indicated
by arrows. Note the amount of
shifts ("hysteresis") as sweep
speed increases!

fig. 1, proportionalist type

fig. 2, velocity and angle
        coder type

fig. 3, movement direction
        coder type

(for more description, see
                      text)

(changes of the dynamic cue, sound motion, its velocity and direction). Therefore neurons were stimulated with electronically produced linear intensity increases in one ear and simultanous decreases in the other. This device simulates horizontal sound movements of constant velocity which was changed systematically (maximally $\pm$ 4o$^{\circ}$ around straight ahead direction corresponding to 4o dB shifts).

The neurons can be classified according to their basical binaural properties but moreover according to their response types to dynamic stimuli as:

1. "Proportionalists" (most common type) code instantanous azimuth angles within a certain working range. The best of these coders indicate the value directly proportionally even during fast movements and independently of movement direction without "hysteresis". The slope of the response increase and decrease is directly proportional to the movement velocity (fig. 1).

2. "Velocity coders" respond with as more activity as faster is the angular velocity, but almost independently of movement direction. Some of these units are activated only when the stimulus sweeps through a certain angle range and sometimes indicate a particular angle by the maximum of excitation independently of velocity and movement direction (fig. 2).

3. "Movement directionalists" respond to motions into one direction with excitation, into the opposite direction with inhibition whereby the amount of the reaction is directly related to velocity (fig. 3). Information on absolute azimuth angles is lost as these units extract the dynamic features movement velocity and direction.

Discussion and conclusion

1. On the brain stem levels studied, neurons with binaural coding of azimuth angles were found as postulated which showed almost no specificity for dynamic cues ("proportionalists"; fig. 1).

2. Although less frequently, some kinds of "feature extracting" neurons were encountered that preferrentially analysed the dynamic cues, sound movement velocity (fig. 2, 3) and direction (fig. 3). Extreme "feature detectors" were not found -; a few rare units, however, only responded significantly with a modulation of a yet stimulus correlated, overall activity (transient excitation and depression; example presented by Neuweiler, this volume). One may ask whether on higher auditory brain levels still higher specializations take place at all.

3. Some but not all dynamic neural response characteristics can already be understood on the base of stationary monaural and binaural input/output functions (comp. with Schlegel, 1977a).

4. The biological relevance of the informations contained in the more or less specialized units' responses for the echolocating bat while hunting in free flight, i.e. for its motor control, might be obvious and should be studied neurophysiologically and behaviourally in more detail.

ALTERATIONS OF AUDITORY RESPONSIVENESS BY THE ACTIVE EMISSION OF ECHOLOCATION SOUNDS IN THE BAT, RHINOLOPHUS FERRUMEQUINUM

Gerd Schuller

During echolocation, the greater horseshoe bat (long CF-FM calls) faces the problem of processing the faint echo during strong acoustic self stimulation by the emitted vocalization. The question arises whether the vocalization acts mainly as a masking stimulus on the echo processing or whether the behavioral state of vocalization prepares the auditory system for the analysis of the echo.

Behavioral findings that the bat only processes Doppler shift information if it is received during vocalization supports the latter hypothesis (Schuller, 1974).

Consequently, single neuron recordings from the inferior colliculus were made in the actively vocalizing greater horseshoe bat. Most neurons with best frequencies between 70 and 86 kHz responded to the electrically elicited vocalizations with response patterns commonly found in collicular neurons. However, if the response to the emitted sound was compared to the response of the same neuron to a pure tone at the CF frequency of the vocalization about 50% of the cells showed differences in the responses to the two stimuli (Fig. 1).

If the vocalization was combined with an artificial echo even more striking effects of vocalization on the collicular neurons were revealed. Some neurons only responded to sinusoidally frequency modulated stimuli (SFM) with synchronized discharges, if they were presented during vocalization. Replacing the vocalization by a pure tone mimicking the CF portion of the vocalization did not result in a synchronized response to the SFM stimulus (Fig. 2).

Other neurons that were stimulated with SFM sounds delayed to the onset of the vocalization only responded with synchronized discharges as long as the echo temporally overlapped with the vocalization.

Using stimulus sets combining the vocalization with frequency-shifted echoes (simulating the Doppler shift compensation system) neurons were found that encoded the threshold for Doppler shift compensation and the frequency shifts relative to the resting frequency by response pattern changes or latency changes respectively (Fig. 3).

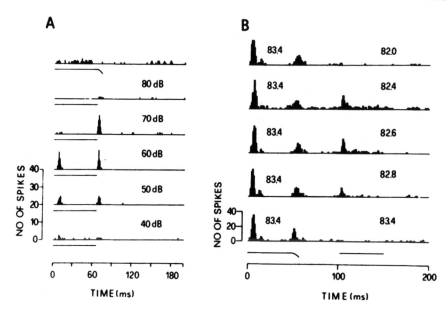

Fig. 1: Responses of 2 collicular neurons to electrically elicited vocalizations and to artificial pure tones. A) Response of the neuron during vocalization (upper histogram) and to artificial pure tones at the frequency of the CF-portion of the echolocation signal at different intensities. B) Response of a neuron during vocalization and subsequent pure tone stimulation at different frequencies (82.0-83.4 kHz) at 80 dB SPL. The CF-frequency of the echolocation signal lies at 83.4 kHz.

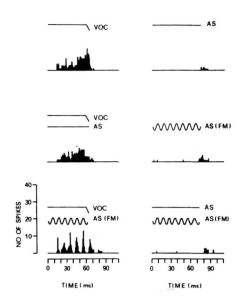

Fig. 2: Discharge patterns of a collicular neuron during vocalization (VOC), to a pure tone at the CF-frequency of the echolocation sound (AS), to a combination of vocalization and the same pure tone (VOC+AS), to a sinusoidally frequency modulated tone (AS(FM)) (carrier frequency: CF-frequency), to a combination of vocalization and the frequency modulated echo (VOC+AS(FM)) and to a combination of a pure tone at the CF-frequency and the frequency modulated echo (AS+AS(FM)). The intensity of the artificial echoes was 80 dB SPL. Note especially the different responses to the two acoustically identical stimulus situations in the lower two graphs.

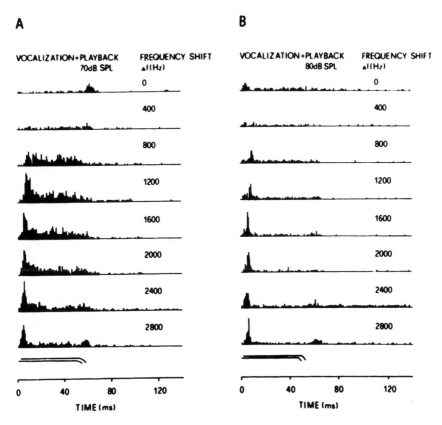

Fig. 3. Responses of 2 collicular neurons to vocalization presented together with frequency-shifted playback echoes, simulating the Doppler shift situation encountered by the
bat during flight towards a target. The echoes were
played back with a delay of 1 ms after the onset of the
vocalization and the frequency shifts are indicated to
the right of each histogram.

These effects of vocalization on the collicular responses to
echoes cannot be explained by mere acoustic interferences between
emitted echolocation pulse and echo, but show that the behavioral
state of active vocalization neuronally affects auditory processing
in collicular neurons when analyzing behaviorally relevant echoes
(SFM signals-wing beat reflections) or Doppler shift information
in echoes.

Supported by the Deutsche Forschungsgemeinschaft, grant number
Schu 390/1,2 and Br 593/2 and Stiftung Volkswagenwerk.

STIMULUS CONTROL OF ECHOLOCATION PULSES IN TURSIOPS TRUNCATUS

Ronald J. Schusterman, Deborah A. Kersting,

and Whitlow W. L. Au

Until recently, it was extremely difficult to determine what cue or set of cues a dolphin used in detecting or differentiating between targets by means of its sonar system. One of the major problems was the ambiguous nature of the echo return in relation to the position of the dolphin during pulse emission and prior to the animal's choice response. In this experiment, we solved the problem by acquiring unambiguous control over a dolphin's position and pulse emission while it was actively engaged in a discriminative echolocation task.

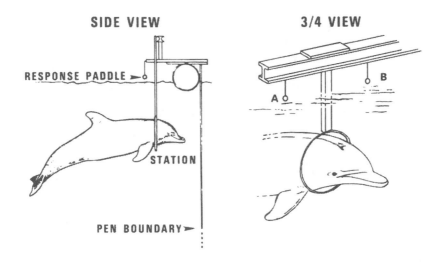

Fig. 1. Diagram of an Atlantic bottlenose dolphin (T. truncatus) maintaining a fixed position in a water-filled plastic hoop. The dolphin remained in the hoop up to its flippers and when given an audio signal, it emitted a burst of echolocating clicks at a target 6 m directly in front of its melon. Following the termination of the last pulse within a burst of pulses, the dolphin backed out of the hoop and reported the presence of a sphere or cylinder by hitting the "A" or "B" response paddle or manipulandum with its rostrum.

An adult male Atlantic bottlenose dolphin (Ekahi), originally signalled by a tone to echolocate and discriminate spheres from cylinders while swimming across his pen, was trained to continue the same task from a fixed position. The animal was first trained to place its head(including blowhole) in a sixteen inch diameter hoop. The hoop was submerged to a depth that placed the dolphin's rostrum directly in line with the center of the target located six meters away (see Fig. 1). When the dolphin maintained the appropriate position in the hoop and emitted no echolocation pulses, the tone cue was activated. The tone signalled the dolphin to echolocate from its position in the hoop. If the animal left the stationary hoop to continue echolocation, the tone was terminated before a choice response was made. If it left the hoop only to surface and make a choice response, the tone remained activated until a response was made. With continued training, control over both the position of the dolphin and the time that it emitted echolocation pulses was exerted in a precise fashion. Control over echolocation pulses was so refined that termination of the tone invariably resulted in the termination of echolocation pulses even though the animal was stationed correctly and the target was present in front of him.

In order to determine if the dolphin was echolocating before the tone was given, a simultaneous measurement of the animal's signals and the tone cue was performed. A B and K 8103 hydrophone was placed one meter behind the target to measure the animal's echolocation signals. The hydrophone signal was amplified by 20 dB and recorded on one channel of an Ampex FR-1300 instrumentation tape recorder. The audio cue was recorded on another channel. Both channels were examined simultaneously using a dual-trace storage oscilloscope in the chop-mode. Analysis of tape recordings made during testing showed that the porpoise did not emit any echolocation signals while stationary in the hoop before the onset of the tone cue.

A second dolphin (Sven) was trained in essentially the same way as we trained Ekahi. The training took approximately the same amount of time (two weeks) and again resulted in controlling not only the postural orientation, but also the signal emissions of an echolocating dolphin in a materials discrimination task.

Supported by the Naval Ocean Systems Center and ONR contract N0014-77-C-0185.

RESPONSE BIAS AND ATTENTION IN DISCRIMINATIVE ECHOLOCATION

BY TURSIOPS TRUNCATUS

Ronald J. Schusterman, Deborah Kersting

and Whitlow W. L. Au

In the Einstellung experiment, stimulus correlated response set or bias can be induced which renders an animal in a maze "functionally blind" to a subsequent simple solution. Until the bias is extinguished, the animal remains insensitive to cues leading to the goal object. Observations on echolocating dolphins suggest that they too may become subject to an Einstellung phenomenon. The present experiment, which grew out of a series of biosonar studies on underwater material discrimination, tested the notion that a response bias acquired in an insolvable discriminative echolocation task will strongly influence the attention of a dolphin on a solvable discriminative echolocation task.

Method. In a two-alternative forced-choice procedure with targets presented on successive trials at a distance of 6 m (see Fig. 3 in Schusterman, this volume), the task of an adult male T. truncatus (Sven) was to differentiate echoes from 17.8 long hollow aluminum and hollow glass targets--each having two different sized outer diamters (OD) and wall thicknesses (see Fig. 1). Targets were submerged in a vertical orientation about 114 cm below the water surface in approximately the same location. Both large and small aluminum cylinders were to be reported on the A or left manipulandum and the large and small glass cylinders were to be reported on the B or right manipulandum. Results from synthesized dolphin-like clicks reflected from these targets in water are shown in Fig. 1. Target reflection characteristics are virtually identical for large aluminum and glass targets (7.62 cm OD), but the differences between small aluminum and glass(3.81 cm OD) are different in terms of the relative strength of the second component of the envelopes of the matched filter output and also with regard to frequency spectra--particularly between 100 and 130 kHz.

In Phase 1, each test session consisted of all four targets being presented 16 times in a quasi-random sequence. In Phase 2 (the Einstellung phase), a test session consisted of either a solvable discrimination problem (small aluminum and glass targets) or an insolvable problem (large aluminum and glass targets). Test sessions in Phase 2 were sequenced in a quasi-random fashion so that the dolphin would have difficulty "predicting" on which test session a solvable or insolvable problem would be given. In Phase 3, the dolphin was given the solvable task on each test session,

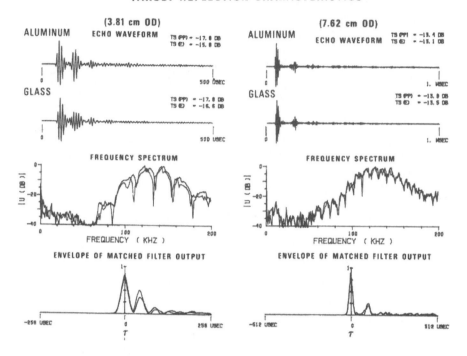

Fig. 1. Results of backscatter measurements for hollow aluminum
cylinders and hollow glass cylinders. Wall thickness of
the smaller OD's was 0.32 cm and that of larger OD's was
0.40 cm. Comparison of echoes from the same sized OD's
can be made on the basis of the time-domain waveforms,
the frequency spectra (superimposition of echoes from
aluminum and glass) and the envelopes of the matched
filter output (superimposition of the echoes from
aluminum and glass).

i.e., the differentiation between small glass and small aluminum.
Following the original experiment, the three phases were replicated
in the same sequence.

Results and Discussion. After 30 test sessions of Phase 1 of
the original experiment, the dolphin perfected the differentiation
between the small aluminum and small glass targets. However, as
predicted from the target reflection characteristics shown in Fig. 1,
the animal did not distinguish between the large aluminum and glass
targets. Instead, the dolphin developed a response bias and consistently reported both large glass and large aluminum on the B
response manipulandum. In Phase 2, response bias not only increased

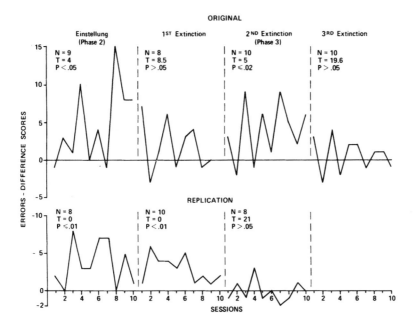

Fig. 2. Comparison of the dolphin's errors (reflecting response bias) during 1st and 2nd halves of each test session during Phases 2 and 3 of the original experiment and the replication. Values on the ordinate were obtained by subtracting errors made during the 2nd half of the test session from errors made during the 1st half of the test session. Positive values indicate response bias during the early part of a test session and 0 values indicate no difference in bias between the 1st and 2nd halves of a test session.

during insolvable large-target sessions, but bias also occurred in the first half of the solvable small-target sessions, decreasing significantly (Wilcoxin matched-pairs signed ranks test) during the second half of the sessions (see Fig. 2).

In Phase 3, when the solvable discrimination was repeatedly given, the initial bias or the Einstellung phenomenon was extinguished. When the entire experiment was replicated, the results were very similar to the first experiment except that the Einstellung phenomenon was extinguished more rapidly. Analysis of click trains from a single test session of Phase 2 showed that the dolphin emitted essentially the same echolocation signals during the solvable discrimination problem regardless of whether or not it had a response bias. The results are interpreted as indicating that persistant spatial responses during an insolvable echolocation task predisposed the dolphin to diminish its attention for listening to previously established distinctive echoes. This interpretation is similar to one suggesting that bats relying on spatial memory pay less attention to echoes in an obstacle avoidance task. (This research was supported by the Naval Ocean Systems Center and ONR contract N00014-77-C-0185.)

THE AUDITORY PATHWAY OF THE GREATER HORSESHOE BAT,

RHINOLOPHUS FERRUMEQUINUM

Hermann Schweizer and Susanne Radtke

Although the central auditory pathway of mammals has been the subject of numerous anatomical studies, only a few anatomical investigations exist on the auditory pathway in echolocating bats. In contrast to this shortage of anatomical knowlege, a lot of neurophysiological data exist from different auditory nuclei. To bridge the gap between anatomical and neurophysiological data we investigated the central auditory pathway of the greater horseshoe bat.

The auditory nuclei were studied with classical histological techniques (Nissl, Klüver-Barrera, Bodian, Golgi). The fiber connections of the inferior colliculus (IC) and the auditory cortex were investigated by using the anterograde and retrograde horse-radish peroxidase (HRP) method. The extension of the auditory cortex and the best frequency range at the HRP-injection sites were determined by using evoked potential recordings.

Auditory centers: No significant differences between the organization of the auditory pathway of this bat species and other mammals were found. This is also true for the composition of the superior olivary complex. In Rhinolophus as well as in Molossus and other investigated bats, a more or less prominent medial superior olivary nucleus was found. The only difference to acoustically non-specialized mammals is, that most of the auditory nuclei are hypertrophied, especially the anteroventral and posteroventral cochlear nucleus, the medial nucleus of the trapezoid body, the lateral superior olivary nucleus, the ventral nucleus of the lateral lemniscus, the central nucleus of the inferior colliculus, the medial geniculate body and the auditory cortex.

The structure and connections of the inferior colliculus (IC): The inferior colliculus of mammals represents an obligatory synapse for lemniscal fibers directly ascending from the lower auditory nuclei. Therefore, the auditory system has no direct pathway to the diencephalic thalamus.

As in other mammals, the IC of Rhinolophus can be divided in three subnuclei: the central nucleus, the pericentral nucleus and the external nucleus. Within the central nucleus a dorsolateral laminated part can be distinguished from a ventromedial non-laminated part. In Golgi-preparations, the same neuron types were found

as in other mammals. A comparison of anatomical and electrophysiological data shows that the frequency band from 10 to about 78 KHz
is represented in the dorsolateral part. The frequencies near the
constant frequency part of the echolocation call, where the audiogram of Rhinolophus is sharply tuned ("filter frequencies"), are
represented in the ventromedial part of the central nucleus of the
IC. The "filter frequencies" are overrepresented in the central
nucleus of the IC as well as in all other lower and higher auditory
centers. Therefore, the distribution of frequencies, within all
auditory centers, directly reflects the distribution of frequencies
along the basilar membrane of the cochlea.

The IC gets direct ipsilateral and/or contralateral inputs in
a tonotopic order from all auditory centers with the exception of
the medial nucleus of the trapezoid body and the medial geniculate
body. Ipsilateral projecting nuclei: dorsal cochlear nucleus,
lateral and ventral nucleus of the trapezoid body, medial superior
olivary nucleus, lateral superior olivary nucleus, dorsal peri-
olivary nuclei, ventral and dorsal nucleus of the lateral lemniscus,
auditory cortex. Contralateral projecting nuclei: dorsal, anteroventral and posteroventral cochlear nucleus, ventral nucleus of
the trapezoid body, lateral superior olivary nucleus, dorsal nucleus
of the lateral lemniscus and the inferior colliculus. These results
show there is a strong convergence of auditory input into the IC;
this points to a very important function of the IC in auditory
processing.

Descending fibers of the IC project to the nuclei of the lateral
lemniscus and to the nuclei of the superior olivary complex. The
ascending fibers project to the medial geniculate body.

Furthermore, the IC is connected to nuclei not belonging to the
classical auditory pathway. These regions are: 1.) The deep layers
of the superior colliculus (they are commonly viewed to have multi-
sensory and motor functions); 2.) The central gray matter (electrical stimulation of the central gray matter elicits vocalization);
3.) The reticular formation of the medulla oblongata (through this
connection an alerting response mediated by auditory signals might
be possible); 4.) The lateral pontine nuclei (these project to the
cerebellum by relaying auditory information to the motor system).

The structure and connections of the auditory cortex: Studies
employing evoked potential recordings and the HRP-technique have
shown that both the auditory cortex and the medial geniculate body
(MGB) of Rhinolophus as the lower auditory centers are vastly
hypertrophied.

The auditory cortex shows the typical six-layered construction
of the mammalian sensory cortex. The anatomical neuron types
correspond to those in other mammals and are layer-specific. The
auditory cortex is tonotopically arranged. The lower frequencies
are located caudally and the higher frequencies rostrally. The

"filter frequencies" are overrepresented (For more detailed information on the tonotopical and functional organization of the auditory cortex, see the contributions of J. Ostwald and N. Suga, this volume).

The auditory cortex gets its acoustical inputs exclusively from the MGB. A detailed analysis of the organization of the MGB and its tonotopical and functional projections is still in progress. But, on the basis of Nissl-stained material, a posterior dorsal cell group can be distinguished from the remaining principal part. Preliminary studies, using relatively large HRP-injections into different frequency regions of the auditory cortex, have shown that in the dorsal cell group as well as in the principal part the entire frequency range of the audiogram is represented. In the principal part, the lower frequencies are localized ventrolaterally and the higher frequencies dorsomedially. The "filter frequencies" are overrepresented and occupy about 50% of the nucleus.

Projections to the auditory cortex also come from other diencephalic regions: from the medial and lateral posterior thalamic nuclei and from the zona incerta. The auditory cortex is reciprocally connected with the contralateral cortex and with the MGB. It also projects back to the IC. Therefore, the auditory cortex could control its own auditory inputs from the diencephalic and mesencephalic levels.

With the techniques used, connections between the auditory cortex and other forebrain centers, the cingular area, the amygdala and a distinct region in the frontal cortex were found. In all of this nuclei, JÜRGENS and PLOOG (1970) were able to elicit vocalization with electrical stimulation in the squirrel monkey. Electrophysiological evidence in Rhinolophus is still lacking, but it is suggested that this forebrain centers might represent vocalization centers, which are presently unknown in bats. (Supported by Deutsche Forschungsgemeinschaft, grant No. Schn.136/6, Br. 593/2.

REFERENCES

JÜRGENS, U., PLOOG, D.: Cerebral representation of vocalization in the squirrel monkey. Exp. Brain Res. 10, 532-554 (1970).

GREY SEAL, HALICHOERUS: ECHOLOCATION NOT DEMONSTRATED

Bill L. Scronce and Sam H. Ridgway

Active echolocation by pinnipeds has been claimed by at least one investigator (Poulter, 1963), but disputed by others (Schusterman, 1967). The grey seal produces clicks. Its underwater hearing is fairly sensitive in the ultrasonic regions important for echolocation in bats and porpoises, and considering the natural environment of the species, we suspected that it used active echolocation.

Our subject was a four-year-old female from Iceland. The seal was trained to wear an opaque elastic band about 15 cm wide that fit over its head blocking vision (Fig. 1). The use of echolocation by this seal was evaluated in two experiments. The first required the seal to retrieve an air-filled plastic ring 20 cm in diameter (D) placed at random locations in a 5 x 1 m section of a 10 m D redwood tank. The second was a forced choice test requiring the seal to detect the flat surface of a 25 cm D styrofoam disk as opposed to the edge of the same disk (-30 dB target strength), placed randomly on either side of a divider.

Without the blindfold the seal's ring retrieval rate was 100% with a latency of 3.8 sec. In 427 blindfolded trials there were 99% correct responses, but latency increased to 6.5 sec. When the 5 x 1 m area, where the ring was randomly placed, was ensonified with echolocation clicks like those employed by dolphins, the seal made 95% correct responses with a latency of 6.1 sec. Head scanning movements were observed on about half the blindfolded trials and click trains were recorded on about 10% of the trials observed.

In the second experiment the blindfolded seal approached the divider and made a choice about target location (right or left) at least 1.5 m away (Fig. 2). Without the blindfold the seal's performance was almost 100%, but with the blindfold correct responses never exceeded 65% and average correct responses for 617 trials was 46%. No click-trains were detected.

The strongest evidence for echolocation was the observation of head scanning movements concurrent with click-trains on some of the blindfolded trials. The animal, however, retrieved the ring just as well when there were no click-trains or head scans. The air-filled ring floating in the water may have been a good passive target for the seal. Since the seal performed the discrimination

Fig. 1. <u>Halichoerus grypus</u> with blindfold in place, vibrissae
         extended.

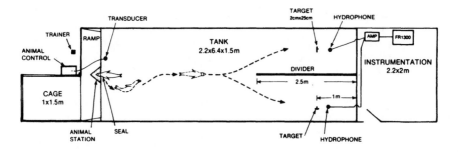

Fig. 2. <u>Halichoerus grypus</u> discrimination experiment setup.

task at chance level, we concluded that the grey seal orients primarily by vision with good capability for passive detection by audition or by tactile senses (vibrissae). The grey seal makes sounds underwater but does not appear to use these sounds for echolocation.

REFERENCES

Poulter, T. C., Sonar signal of the sea lion, Science 139:753.

Schusterman, R. J. 1967, Perception and determinants of underwater vocalization in the California sea lion. In: R. G. Busnel (Ed.), Animal Sonar Systems, Vol. 1. Laboratoire de Physiologie Acoustique, Jouy-en-Josas, France, 535-617.

THE ROLE OF THE ANTERIOR AND POSTERIOR CRICOTHYROID MUSCLES IN THE PRODUCTION OF ECHOLOCATIVE PULSES BY MORMOOPIDAE

Roderick A. Suthers and Gary E. Durrant

Pteronotus parnellii in the family Mormoopidae emits pulses having a long initial CF portion followed by a short downward FM sweep. Most of the acoustic energy is in the second harmonic at 60 kHz, but the fundamental and higher harmonics can also be introduced when greater bandwidth is needed (Fig. 1). Griffiths (1978) reported that the cricothyroid muscle (CTL) of Mormoopidae has two distinct bodies--an anterior longitudinal portion (aCTM) and a posterior transverse portion (pCTM). The latter attaches to the flexible lateral walls of the laryngeal cavity. On the basis of its anatomy, Griffiths suggested that it might enable the bat to tune its larynx, in the manner of a variable Helmholtz resonating amplifier, by altering the cross-sectional area of the laryngeal cavity anterior to a larger post-laryngeal chamber formed by the enlarged trachea. We have examined the effect of a light gas mixture (20% oxygen, 80% helium) on the harmonic content and frequency structure of orientation pulses emitted before and after selective denervation of particular parts of the CTM in an effort to determine if the pCTM is responsible for varying the harmonic content of the emitted pulse by changing the vocal tract resonance as suggested by Griffiths. By increasing the speed of sound, without altering the vibration of the laryngeal generator or changing the dimensions of the airway, a light gas mixture enables one to evaluate the extent to which the emitted frequencies depend on resonance or filtering by the vocal tract.

Denervation of aCTM. Bilateral denervation of the aCTM causes no pronounced change in harmonic emphasis. Pulse frequency rises slightly during the first few msec before becoming CF and the frequency of the CF portion of the second harmonic is about 2 kHz below normal. Light gas mixtures suppress the second harmonic and enhance the fundamental and third harmonic as they do in the case of a normal bat. The aCTM appears not to play a major essential role in controlling the frequency composition of emitted pulses.

Denervation of pCTM. Bilateral denervation of the pCTM also results in a 2 to 3 kHz drop in the frequency of the second harmonic (Fig. 2). The terminal FM sweep begins earlier in the pulse and proceeds at a more gradual rate than is normal. The intensity of the third harmonic is increased to approximately equal that of the second harmonic. The fundamental, fourth and sometimes higher harmonics, are usually also present at a lower intensity. A light gas mixture shifts the emphasis to the fundamental and third

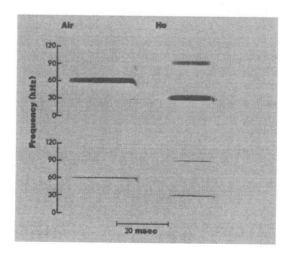

Fig. 1. Typical orientation pulses produced by an intact bat
in a plastic bag filled with air or with a light gas
mixture (He). Wide band sonagrams are above narrow
band sonagrams.

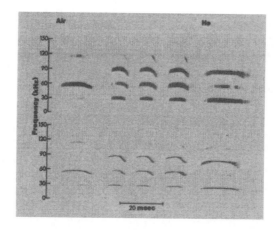

Fig. 2. Pulses emitted in air and light gas mixture after
bilateral denervation of the pCTM.

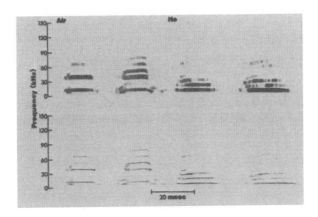

Fig. 3. Pulses emitted in air and light gas mixture after
         bilateral denervation of the pCTM and aCTM.

harmonic, but the second and fourth harmonics are still present.

    Denervation of aCTM and pCTM. Bilateral denervation of both
aCTM and pCTM causes a dramatic drop in generator frequency (Fig. 3).
The fundamental drops to 10 or 12 kHz and as many as 9 or 10
harmonics may be present. Harmonics in the range between about 40
and 65 kHz often tend to be emphasized in air, but suppressed in
the light gas mixture, indicating a significant amount of filtering
by the vocal cavities remains.

    Conclusions. The aCTM and pCTM together determine the fre-
quency of the vocal generator. If either portion is paralyzed, the
remaining part can largely compensate for its absence. The aCTM
is particularly important in establishing the frequency at the
beginning of the pulse, while the pCTM is also important in timing
the start and controlling the rate of the terminal FM sweep. The
pCTM may play a role in controlling the harmonic content of the
emitted pulse, possibly by varying the tuning of the vocal tract.
However, a significant amount of vocal tract tuning remains after
denervation of both the a CTM and pCTM, indicating the existence of
a second independent filter.

(Supported by NSF grant BNS 76-01716).

CODING OF SINUSOIDALLY FREQUENCY-MODULATED SIGNALS BY SINGLE COCHLEAR NUCLEUS NEURONS OF RHINOLOPHUS FERRUMEQUINUM

Marianne Vater

The Greater Horseshoe bat, Rhinolophus ferrumequinum, emitts a long constant-frequency (CF-) component of about 83 kHz during echo-location. The echoes are Doppler-shifted upward in frequency due to the realative movement between bat and target. The bat compensates for these deviations in the echo-frequency by lowering the frequency of the emitted call so that the echo-frequency remains constant within a narrow frequency band (Schnitzler, H.-U., 1968; Schuller et al., 1974). The wing beat of prey insects results in periodic frequency and amplitude modulations of this carrier frequency which can be used by the bats as clues for detection and probably identification of moving prey objects. Neurophysiological investigations of the inferior colliculus (IC) and auditory cortex of Rhinolophus (Schuller, G., 1979; Ostwald, J., this volume) showed that the sharply tuned filterneurons around 83 kHz specifically analyse the information contained in periodic modulations.

In order to examine the neuronal encoding characteristics of these complex time-varying signals within the ascending auditory pathway of Rhinolophus, recordings were made from single cochlear nucleus (CN) neurons. The sample comprises neurons from all subdivisions of the CN-complex, localized with Alcainblue spots. Various parameters of sinusoidally frequency-modulated (SFM-) signals were presented.

The neurons synchronize their discharge activity to the period of the SFM-stimulus. Tonic neurons were found that faithfully reproduced the time course of the periodic modulation. It can therefore be expected that signals with more complex temporal patterns, as for example generated by the wing beat of prey insects are also reproduced with only little distortion and relayed as input information to higher auditory centers. Other tonic cells asymmetrically enhanced the upward or downward component of the modulation cycle in their discharge pattern. Phasic-on neurons did not preserve the stimulus time structure: they responded with a transient discharge activity to distinct portions of the modulation cycle.

The modulation height of the periodic modulation of the echo-frequency depends on the size of the prey insect. CN-neurons code a wide range of modulation heights thereby fulfilling the criteria for processing echo-information of a variety of prey sizes. The threshold for modulation height was found to be especially low in CN-filterneurons (minimum + 20 to + 50 Hz) as compared with neurons of other frequency ranges. These minimum modulation heights are comparable to the values found by Schuller, G. (1979) for IC-neurons, therefore no further neuronal interaction is required to account for this response property of the higher auditory neurons. Rather, the narrow bandwidth and the steep tuning curve flanks of filterneurons, which are the result of specialized mechanical properties of the cochlea (Bruns, V., 1976a, b) seem to be responsible for this response characteristic. Other factors governing the response to SFM are discussed later.

SFM-stimuli centered at the best frequency of a neuron are not necessarily the optimal stimulus. Shifting the carrier frequency towards the tuning curve flanks can lead to a better synchronization and to a lowering of the modulation threshold.

The echo intensity depends on the distance between bat and target. CN-neurons clearly synchronize their discharges to the modulation cycle over a wide range of stimulus intensities.

The response to SFM-stimuli of different modulation frequencies correlates with the neurons' response patterns to CF-stimuli and therefore depends on the localization within the different subdivisions of the CN. Primary-like neurons of the anteroventral CN (Fig. 1 A) are able to synchronize their discharges to modulation frequencies up to 900 Hz. Functions, relating the magnitude of the neurons response to modulation frequency typically possess an optimum around 400 Hz. The stimulus coding in the time domain therefore far exceeds the frequencies of prey insect wing beat (i.e. between 30 and 100 Hz). In contrast, build up and complex neurons of the dorsal CN, even if they are sharply tuned, were unable to follow the periodic modulations in the modulation frequency range tested (minimum modulation frequency 20 Hz). Their response patterns to SFM-stimuli resembles the patterns to CF-stimuli. Therefore, the structure of the synaptic connections within the subnuclei of the CN also plays a role in the neurons' responses to periodic stimuli. Phasic-on neurons in the dorsal CN (Fig. 1 B) typically lock only to modulation frequencies below 300-400 Hz with an optimum response below 100 Hz, a range largely restricted to the biologically important parameters. These response properties were also reported for IC-neurons of Rhinolophus (Schuller, G., 1979).

These results show that this peripheral stage of the ascending auditory pathway can not be treated as a uniform population of neurons simply relaying information to higher auditory centers for further processing, but that the extraction of biologically important parameters has already begun.

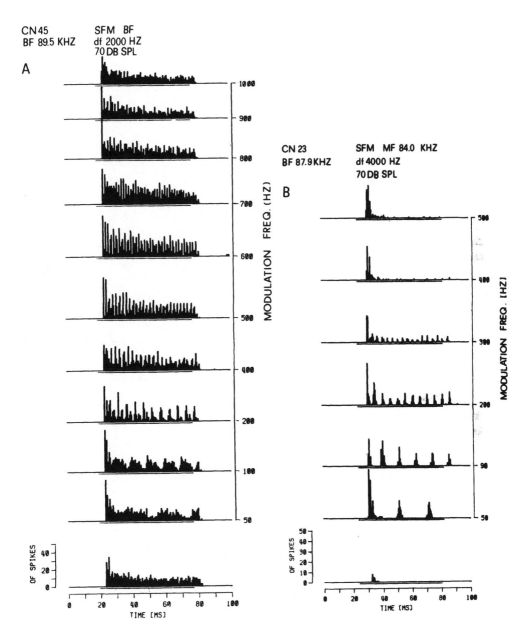

Fig. 1. Response of anteroventral (A) and dorsal (B) CN-neurons to SFM-stimuli of varying modulation frequency. Lower histogramms: response to CF-signals at best frequency (BF; A) or to CF-signals at the carrier frequency (MF; B) of the SFM. df: modulation height; stimulus intensity as indicated. Binwidth 0.5 ms, 50 stimuli. bars: duration.

SINGLE UNIT RESPONSES TO FREQUENCY-MODULATED SOUNDS AND SIGNAL-NOISE COMBINATIONS. A COMPARATIVE STUDY OF THE INFERIOR COLLICULUS OF MOLOSSID BATS

Marianne Vater and Peter Schlegel

Recordings were made from single inferior colliculus neurons of two closely related bat species, Molossus ater and Molossus molossus. Their short CF-FM echolocation calls differ only in frequency range.

The audiograms of both species derived from evoked potential measurements and threshold distributions of single neurons are broadly tuned with maximum auditory sensitivity reached at different frequency ranges according to the different spectral content of the orientation calls. Tuning curves, Q_{10} dB values, response patterns and spike count functions of single units are very similar in the samples obtained from both molossid species and closely resemble data from bats using FM orientation calls (Pollak et al., 1978). Contrary to long CF-FM bats such as Rhinolophus ferrumequinum, no conspicuous neuronal adaptations to the short CF component were found in molossids.

However, the use of FM stimuli revealed a high degree of specialization to the FM component of the orientation call. Artificial FM pulses of different durations, modulation heights and directions centered at the neurons' best frequency were tested. In about 30% of the neurons, thresholds to FM pulses mimicking the natural FM part of the echolocation call in sweep rate, were lower than thresholds to pure tone pulses of equal duration (2 ms), resulting in a better detectability of FM.

Response patterns of single units depended on the frequency time course of the stimulus. Neuronal response activity to slowly sweeping FM pulses could be greater than to any other stimulus configuration. Stabilization of initial latency was frequently observed with fast sweeping FM, resulting in increased accuracy of time information.

Threshold signal/noise ratios, measured by masking the neuron's response to the signal by noise, were usually equal for CF and FM stimuli or lower for the FM signal as compared to the CF signal. This increased detectability of FM sweeps in noise is advantageous for the bat usually hunting under nonideal acoustical conditions.

ASCENDING AUDITORY PATHWAYS IN THE BRAIN STEM OF THE BAT,

PTERONOTUS PARNELLII

J. M. Zook and J. H. Casseday

Although mammalian echolocation has been extensively studied in recent years, little is known concerning the connections of neural centers which might be involved in the special auditory functions of echolocation. We have mapped the ascending auditory pathways from cochlear nucleus in the bat, Pteronotus parnellii parnellii (Gray), using techniques for tracing axonally trans-ported molecules (^{3}H leucine or horseradish peroxidase) in the anterograde or retrograde directions. Our principal observations concern (1) direct pathways from cochlear nucleus to the central nucleus of the inferior colliculus and (2) an indirect pathway which relays within the lateral lemniscus enroute from cochlear nucleus to the central nucleus of the inferior colliculus.

The results show first that the direct projections of the three subdivisions of the cochlear nucleus, the anteroventral, posteroventral and dorsal cochlear nuclei, do not completely over-lap within the central nucleus of the inferior colliculus. The anteroventral cochlear nucleus projects only to the ventral and lateral parts of the central nucleus. From the posteroventral cochlear nucleus, the region of octopus cells projects to an even more limited area of the ventral part of the central nucleus of the inferior colliculus. Only the dorsal cochlear nucleus pro-jects to all parts of the central nucleus of the inferior col-liculus, but this projection is most dense in the dorsal and lat-eral aspects of the central nucleus. We conclude that each of the major subdivisions of the cochlear nucleus has a specific target in the central nucleus of the inferior colliculus but that there is partial overlap of one target with another. Additional obser-vations reveal topographic projections from the anteroventral, posteroventral and dorsal cochlear nuclei to their respective targets.

In addition to the direct pathway from the subdivisions of the cochlear nucleus to the central nucleus of the inferior col-liculus, Pteronotus, as other mammals, has several indirect path-ways which relay in nuclei in the superior olivary complex and the lateral lemniscus before terminating in the central nucleus of the inferior colliculus. One of these indirect pathways may be unique to Pteronotus in that the pathway relays within the lateral lemniscus at a large nucleus which has no homologue in mammals that do not echolocate. This "intermediate" nucleus of the lat-eral lemniscus receives its input primarily from the anteroventral cochlear nucleus and sends its main projections in turn to the central nucleus of the inferior colliculus. It is quite possible that this indirect pathway from cochlear nucleus to the central nucleus of the inferior colliculus is an elaboration unique to echolocating bats.

(Supported by NSF grant BNS 77-°6796)

Chapter VIII
General Bibliography

- Bibliography on echolocation in bats.
     compiled by G. Neuweiler

- Bibliography on echolocation in odontocetes.
     compiled by P. Nachtigall

- Bibliography on echolocation in non-bat non-cetacean
  species.
     compiled by E. Gould

BIBLIOGRAPHY ON ECHOLOCATION IN BATS

(Literature cited: articles published from 1965 to 1. Febr. 1979.
Russian literature is not included.)

1. Aidley, D. J. Echo intensity in range estimation by bats. Nature, 224, 1330-1331. 1969.
2. Altes, R. A. Optimum waveforms for sonar velocity discrimination, in: Proc. IEEE, 59, 1615-1617. 1971.
3. ―――― Suppression of radar clutter and multipath effects for wide-band signals. IEEE Transactions on Information Theory, IT-17, 344-346. 1971.
4. ―――― Methods of wideband signal design for radar and sonar systems. Federal Clearinghouse Number AD 732 494. 1971.
5. ―――― Some invariance properties of the wide-band ambiguity function. J. Acoust. Soc. Am., 53, 1154-1160. 1973.
6. ―――― Study of animal systems with application to sonar. Progress Report No. ESL PR 144, Electromagnetic Systems Labs., Sunnyvale, Cal. 1974.
7. ―――― Mechanism for aural pulse compression in mammals. J. Acoust. Soc. Am., 57, 513-515. 1975.
8. ―――― Sonar for generalized target description and its similarity to animal echolocation systems. J. Acoust. Soc. Am., 59, 97-105. 1976.
9. ―――― The Fourier-Mellin transform and mammalian hearing. J. Acoust. Soc. Am., 63, 174-183. 1978.
10. ―――― Angle estimation and binaural processing in animal echolocation. J. Acoust. Soc. Am., 63, 155-173. 1978.
11. ―――― Further development and new concept for bionic sonars, Vols. II and III. Report OC-R-78-A 004-1, ORINCON Corp., 3366 N. Torrey Pines Ct., La Jolla, CA. 92037. 1978.
12. ――――, and Faust, W. I. Further development and new concept for bionic sonar, Vol. I. Report OC-R-78-A 004-1, ORINCON Corp., 3366 N. Torrey Pines Ct., La Jolla, CA. 92037. 1978.
13. ――――, and Reese, W. D. Doppler-tolerant classification of distributed targets - a bionic sonar. IEEE Transactions Aerospace and Electronic Systems, Vol. AES-11 (5), 708-724. 1975.
14. ――――, and Skinner, D. P. Sonar velocity resolution with a linear-period-modulated pulse. J. Acoust. Soc. Am., 61, 1019-1030. 1977.
15. ――――, and Titlebaum, E. L. Bat signals as optimally Doppler tolerant waveforms. J. Acoust. Soc. Am., 48, 1014-1020. 1970.
16. ――――, ―――― Graphical derivations of radar, sonar, and communication signals. IEEE Transactions Aerospace and Electronic Systems, Vol. AES-11 (1), 38-44. 1975.
17. Anderson, G. M. A model for the bat versus moth pursuit-evasion problem. J. Acoust. Soc. Am., Suppl. No. 1, S88. 1978.

18. Bach Andersen, B., and Miller, L. A.   A portable ultrasonic de-
    tection system for recording bat cries in the field. J. Mammal.,
    58, 226-229. 1977.
19. Barclay, R. M. R.   Vocal communication and social behavior of
    the little brown bat, Myotis lucifugus (Chiroptera: Vespertilioni-
    dae). M. Sc. Thesis, Dept. of Biology, Carleton University, Ottawa,
    Canada, 1978.
20. Baron, G.   Morphologie comparative des relais auditifs chez les
    chiroptères. Dissertation, Montreal, 1972.
21. ———— Differential phylogenetic development of the acoustic
    nuclei among chiroptera. Brain Behav. Evol., 9, 7-40.   1974.
22. Beuter, K. J.   Systemtheoretische Untersuchungen zur Echoortung
    der Fledermäuse. Dissertation, Universität Tübingen, 1976.
23. ———— Maskierung von konstantfrequenten Echoortungslauten mit
    Bandpaßrauschen bei der Fledermaus Rhinolophus ferrumequinum.
    Verh. Dtsch. Zool. Ges., 1978, 195, G. Fischer Verl., Stuttgart,
    1978.
24. ———— Optimalempfängertheorie und Informationsverarbeitung im
    Echoortungssystem der Fledermäuse, in: "Kybernetik 77," Oldenburg-
    Verlag, München, 1978.
25. Bradbury, J. W.   Target discrimination by the echolocating bat
    Vampyrum spectrum. J. Exp. Zool., 173, 23-46.   1970.
26. ————, and Vehrencamp, S. L.   Social organization and foraging
    in emballonurid bats. I. Field studies. Behav. Ecol. Sociobiol.,
    1, 337-381.   1976.
27. Brosset, A.   "La Biologie des Chiroptères," Masson et Cie., Paris,
    1966.
28. Brown, A. M.   An investigation of the cochlear microphonic res-
    ponse of two species of echolocating bats, Rousettus aegyptiacus
    (Geoffroy) and Pipistrellus pipistrellus. J. comp. Physiol., 83,
    407-413.   1973.
29. ————, and Pye, J. D.   Auditory sensitivity at high frequencies
    in animals. Adv. Comp. Physiol. Biochem., 6, 1-73.   1975.
30. Brown, P. E.   Vocal communication and the development of hearing
    in the pallid bat, Antrozous pallidus. Ph. D. Thesis, University
    of California Los Angeles, 1973.
31. ———— Vocal communication in the pallid bat, Antrozous palli-
    dus. Z. Tierpsychol., 41, 34-54.   1976.
32. ————, Grinnell, A. D., and Harrison, J. B.   Developing of
    hearing in pallid bat, Antrozous pallidus. J. comp. Physiol. A,
    126, 169-182.   1978.
33. Bruns, V.   Die Anpassung des Hörsystems der Hufeisennase an die
    Echoortung mit konstantfrequenten Ortungslauten. Verh. Dtsch. Zool.
    Ges., 1976, 268, G. Fischer Verl., Stuttgart, 1976.
34. ———— Peripheral auditory tuning for fine frequency analysis
    by the CF-FM bat, Rhinolophus ferrumequinum. I. Mechanical special-
    izations of the cochlea. J. comp. Physiol. A, 106, 77-86.   1976.
35. ———— Peripheral auditory tuning for fine frequency analysis
    of the CF-FM bat, Rhinolophus ferrumequinum. II. Frequency mapping
    in the cochlea. J. comp. Physiol. A, 106, 87-97.   1976.

36. Bruns, V. Die Feinstruktur der für die Hydrodynamik wesentlichen Elemente der Cochlea der Großen Hufeisennase, Rhinolophus ferrumequinum. Verh. Dtsch. Zool. Ges., 1978, 154, G. Fischer Verl., Stuttgart, 1978.

37. ———, Neuweiler, G., and Suga, N. Specialized frequency processing in the auditory system of the echolocating horseshoe-bat. Exp. Brain Res., 23, Suppl., 30. 1975.

38. Burikova, N. N. Efferent links in the cerebral auditory cortex of the bat Myotis oxygnathus. J. Evol. Biochem. Physiol., 7, 450-455. 1971.

39. Busnel, R. G., ed. "Animal Sonar Systems," Vol. I and Vol. II. Lab. Physiol. Acoust., C. N. R. Z., Jouy-en-Josas, France, 1967.

40. Cahlander, D. A., McCue, J. J. G., and Webster, F. A. The determination of distance by echolocating bats. Nature, 201, 544-546. 1965.

41. Dalland, J. I. Auditory thresholds in bats: a behavioral technique. J. Aud. Res., 5, 95-108. 1965.

42. ——— Hearing sensitivity in bats. Science, 150, 1185-1186. 1965.

43. ———, Vernon, J. A., and Peterson, E. A. Hearing and cochlear microphonic potentials in the bat Eptesicus fuscus. J. Neurophysiol., 30, 697-709. 1967.

44. David, A. Can young bats communicate with their parents at a distance? J. Bombay Natur. Hist. Soc., 65, 210. 1968.

45. Dunning, D. C., and Roeder, K. D. Moth sounds and the insect catching behaviour of bats. Science, 147, 173-174. 1965.

46. Dziedzic, A. Quelques performances des systèmes de detection par echos des chauve-souris et de delphinidae. Rev. Acoust., 1, 23-28. 1968.

47. Engelstätter, R. Einzelzellantworten aus dem Colliculus inferior von Rhinolophus ferrumequinum auf Reizung mit Rauschen unterschiedlicher Bandbreite und Kombinationen von Rauschen und Reintönen. Diplomarbeit, Fachbereich Biologie, Universität Frankfurt, 1978.

48. Escudie, B. Etat actuel des travaux sur les "sonars biologique", comparaison des propriétés de ces systemes avec les systèmes artificiels. I. C. P. I. de Lyon, Report TS 72/02. 1972.

49. ———, and Hellion, A. Comparison entre certains signaux optimaux a grand Bt et ceux utilisés par les chauve-souris, in: C. R. du contrat DRME 72419, "Modèle de reception acoustique biologique ...", 1973.

50. Fairbanks, J. T. Detection and triggering device for research on bat-echolocation systems. Mass. Inst. Techn. Res. Lab. Electr. Quart. Prog. Rep., 86, 337-338. 1967.

51. Feng, A. S., Simmons, J. A., and Kick, S. A. Echo detection and target-ranging neurons in the auditory system of the bat Eptesicus fuscus. Science, 202, 645-648. 1978.

52. Fenton, M. B. The role of echolocation in the evolution of bats. American Naturalist, 108, 386-388. 1974.

53. ——— Acuity of echolocation in Collocalia hirundinacea (Aves: Apodidae), with comments on the distribution of echolocating swiftlets and Molossid bats. Biotropica, 7, 1-7. 1975.

54. Fenton, M. B. Observations on the biology of some Rhodesian bats, including a key to the Chiroptera of Rhodesia. Life Sciences Contribution, Royal Ontario Museum, No. 104, 1-27. 1975.

55. ———, Boyle, N. G. H., Harrison, T. M., and Oxley, D. J. Activity patterns, habitat use and prey selection by some African insectivorous bats. Biotropica, 9, 73-85. 1977.

56. ———, Cumming, D. H. M., and Oxley, D. J. Prey of bat hawks and availability of bats. Condor, 79. 495-497. 1977.

57. ———, Jacobson, S. L., and Stone, R. N. An automatic ultrasonic sensing system for monitoring the activity of some bats. Can. J. Zool., 51, 291-299. 1973.

58. Firbas, W. The number of neurons in the cochlea of some bats. J. Mammal., 51, 809-810. 1970.

59. ——— Über anatomische Anpassungen des Hörorgans an die Aufnahme hoher Frequenzen. (Eine vergleichend-anatomische Untersuchung an Fledermäusen.) Monatsschr. Ohrheilk. Laryng. Rhin., 106, 105-156. 1972.

60. ———, and Einzinger, H. Über das Ganglion spirale der Chiroptera. Z. Säugetierkunde, 37, 321-326. 1972.

61. ———, Welleschik, B. Über die Verteilung der Acetylcholinesterase-Aktivität im Cortischen Organ von Fledermäusen. Acta Oto-Laryng., 70, 329-335. 1970.

62. Flieger, E., and Schnitzler, H.-U. Ortungsleistungen der Fledermaus Rhinolophus ferrumequinum bei ein- und beidseitiger Ohrverstopfung. J. comp. Physiol., 82, 93-102. 1973.

63. Friend, J. H., Suga, N., and Suthers, R. A. Neural responses in the inferior colliculus of echolocating bats to artificial orientation sounds and echoes. J. Cell. Comp. Physiol., 67, 319-332. 1966.

64. Fullard, J. H., and Fenton, M. B. Acoustic and behavioural analysis of the sounds produced by some species of Neoarctic Arctiidae (Lepidoptera). Can. J. Zool., 55, 1213-1224. 1977.

65. Funakoshi, K., and Uchida, T. A. Studies on the physiological and ecological adaptations of temperate insectivorous bats. I. Feeding activities in the Japanese long-fingered bats. Japanese J. Ecol., 25, 217-234. 1975.

66. Glaser, W. Eine systemtheoretische Interpretation der Fledermausortung. Studia Biophysica, 27, 103-110. 1971.

67. ——— Zur Fledermausortung aus dem Gesichtspunkt der Theorie gestörter Systeme. Zool. Jb. Physiol., 76, 209-229. 1971.

68. ——— Zur Hypothese des Optimalempfangs bei der Fledermausortung. J. comp. Physiol., 94, 227-248. 1974.

69. Goldman, L. J., Henson, O. W. Prey recognition and selection by the constant frequency bat, Pteronotus p. parnellii. Behav. Ecol. Sociobiol., 2, 411-420. 1977.

70. Gould, E. "Echolocation and communication in bats." Southern Methodist University Press, Dallas, Texas 75222, 1970.

71. ——— Individual recognition by ultrasonic communication between mother and infant bats (Myotis), in: "Animal Orientation and Navigation," S. R. Galler, K. Schmidt-König, G. J. Jacobs, R. E. Belleville, eds. NASA Press, Virginia, USA, 1970.

72. Gould, E. Neonatal vocalizations in bats of eight genera. J. Mammal., 56, 15-29. 1975.

73. ———— Experimental studies on the ontogeny of ultrasonic vocalizations in bats. Dev. Psychobiol., 8, 333-346. 1975.

74. ————, Lemkau, P. V., and Hume, J. C. "Ontogeny of Echolocation in Bats." J. Hopkins Univ., Dept. Mental Hygiene, Baltimore, MD. 21218, 1972. 105 pp.

75. Greguss, P. Bioholography - a new model of information processing. Nature, 219, 482. 1968.

76. ———— Variable holography. Laser in der Forschung, (2), 43-46, (3), 57-60. 1969.

77. Griffin, D. R. Discriminative echolocation by bats, in: "Animal Sonar Systems," R. G. Busnel, ed., pp. 273-306. Lab. Physiol. Acoust., C. N. R. Z., Jouy-en-Josas, France, 1967.

78. ———— Migration and homing of bats, in: "Biology of Bats," Vol. I, W. A. Wimsatt, ed. Academic Press, New York, 1970.

79. ———— The importance of atmospheric attenuation for the echolocation of bats. Anim. Behav., 19, 55-61. 1971.

80. ———— Comments on animal sonar symposium. J. Acoust. Soc. Am., 54, 137-138. 1973.

81. ———— Echolocation, in: "Basic Mechanisms in Hearing," A. Møller, ed. Academic Press, New York, 1973.

82. ———— "The Sensory Physiology of Animal Orientation." Harvey Lectures Series 71, Academic Press, New York, 1978.

83. ————, Friend, J. H., and Webster, F. A. Target discrimination by echolocation in bats. J. Exp. Zool., 158, 155-168. 1965.

84. ————, and Simmons, J. A. Echolocation of insects by horseshoebats. Nature, 250, 731-732. 1974.

85. Grinnell, A. D. Mechanisms of overcoming interference in echolocating animals, in: "Animal Sonar Systems," R. G. Busnel, ed., pp. 451-481. Lab. Physiol. Acoust., C. N. R. Z., Jouy-en-Josas, France, 1967.

86. ———— Comparative physiology of hearing. Ann. Rev. Physiol., 31, 545-580. 1969.

87. ———— Comparative auditory neurophysiology of neotropical bats employing different echolocation signals. Z. vergl. Physiol., 68, 117-153. 1970.

88. ———— Neural processing mechanisms in echolocating bats, correlated with differences in emitted sounds. J. Acoust. Soc. Am., 54, 147-156. 1973.

89. ———— Rebound excitation (off-responses) following non neural suppression in the cochleas of echolocating bats. J. comp. Physiol., 82, 172-194. 1973.

90. ————, and Brown, P. Long-latency "subthreshold" collicular responses to the constant-frequency components emitted by a bat. Science, 202, 996-999. 1978.

91. ————, and Grinnell, V. S. Neural correlates of vertical localization by echolocating bats. J. Physiol., 181, 830-851. 1965.

92. ————, and Hagiwara, S. Adaptation of the auditory system for echolocation: studies of New Guinea bats. Z. vergl. Physiol., 76, 41-81. 1972.

93. Grinnell, A. D., and Hagiwara, S. Studies of auditory neurophysiology in non-echolocating bats, and adaptations for echolocation in one genus Rousettus. Z. vergl. Physiol., 76, 82-96.
1972.

94. ————, and Schnitzler, H.-U. Directional sensitivity of echolocation in the horseshoe bat Rhinolophus ferrumequinum. II.
Behavioral directionality of hearing. J. comp. Physiol., 116,
63-76. 1977.

95. Gunier, W. Experimental homing of gray bats to a maternity
colony in a Missouri barn. Am. Midl. Nat., 86, 502-506. 1971.

96. Habersetzer, J. Ortungslaute der Mausohrfledermaus (Myotis
myotis) in verschiedenen Ortungssituationen. Verh. Dtsch. Zool.
Ges., 1978, 196, G. Fischer Verl., Stuttgart, 1978.

97. Hall, J. G. The cochlea and the cochlear nuclei in the bat.
Acta Otolaryngol., 67, 490-500. 1969.

98. Halls, J. A. T. Radar studies of the bat sonar. Proc. 4th. Int.
Bat Res. Conf. Kenya Literat. Bureau, Nairobi (eds. Olembo als.),
137-142. 1978.

99. Harrison, J. B. Temperature effects on responses in the auditory
system of the little brown bat Myotis l. lucifugus. Physiol. Zool.,
38, 34-48. 1965.

100. Henson, M. M. Unusual nerve-fibre distribution in the cochlea
of the bat Pteronotus p. parnellii (Gray). J. Acoust. Soc. Am.,
53, 1739-1740. 1973.

101. ———— The basilar membrane of the bat, Pteronotus p. parnellii. American J. Anat., 153, 143-158. 1978.

102. ———— Structural changes in the inner ear of the bat,
Pteronotus p. parnellii after overstimulation with constant
frequency signals. Amer. J. Anat., 190, 610 (Abst.). 1978.

103. ————, and Lay, D. M. Scanning electron microscopy of the
basilar membrane. Anat. Rec., 184, 425 (Abst.). 1976.

104. Henson, O. W. The activity and function of the middle-ear
muscles in echolocating bats. J. Physiol., 180, 871-887. 1965.

105. ———— Fluctuations of middle ear air pressure in bats.
Anat. Rec., 151, 456. 1965.

106. ———— Comparative physiology of middle ear muscle activity
during echolocation in bat. Amer. Zool., 6, 603. 1966.

107. ———— Auditory sensitivity in Molossidae. Anat. Rec., 157,
363-364. 1967.

108. ———— The perception and analysis of biosonar signals by
bats, in: "Animal Sonar Systems," R. G. Busnel, ed., II., pp. 949-
1003. Lab. Physiol. Acoust., C.N.R.Z., Jouy-en-Josas, France, 1967.

109. ———— Neuroanatomical and physiological correlates of acoustic
orientation. Anat. Rec., 163, 305 (Abst.). 1969.

110. ———— The central nervous system of Chiroptera, in: "Biology
of Bats," W. A. Wimsatt, ed., chap. 2, 57-152. Academic Press,
New York, 1970.

111. ———— The ear and audition, in: "Biology of Bats," W. A.
Wimsatt, ed., vol. II, chap. 4. Academic Press, New York, 1971.

112. Henson, O. W.   Comparative anatomy of the middle ear, in:
     "Handb. Sens. Physiol.", W. D. Keidel and W. D. Neff, eds.,
     vol. V, pp. 39-100. Springer Verlag, Berlin, Heidelberg, New York,
     1974.
113. ———, Eager, R. P., and Henson, M. M.   Tecto-cerebellar pro-
     jections in the bat Tadarida brasiliensis. Anat. Rec., 160, 473-
     474 (Abst.). 1968.
114. ———, and Henson, M. M.   Middle ear muscle contractions and
     their relation to pulse- and echo-evoked potentials in the bat,
     in: "Animal Orientation and Navigation," S. R. Galler et al., eds.,
     NASA Press, Washington, 1972.
115. ———, and Pollak, G. D.   A technique for chronic implanta-
     tion of electrodes in the cochlea of bats. Physiol. Behav. 8,
     1185-1187. 1972.
116. ———, ———, and Novick, A.   Continuous monitoring of
     cochlear potentials and middle ear muscle activity in echolocating
     bats. Anat. Rec., 172. 328. 1972.
117. Hinchcliffe, R., and Pye, A.   The cochlea in Chiroptera: a
     quantitative approach. Int. Audiol., 7, 259-266. 1968.
118. ———, ———   Variations in the middle ear of the mammalia.
     J. Zool., 157, 277-288. 1969.
119. Hooper, J. H. D.   Potential use of a portable ultrasonic receiver
     for the field identification of flying bats. Ultrasonics, 7, 177-
     181. 1969.
120. Howell, D. J.   Acoustic behavior and feeding in Glossophagine
     bats. J. Mammal., 55, 293-308. 1974.
121. Jen, P. H.-S.   Coding of the directional information of single
     neurons in the superior olivary complex of echolocating bats.
     Ph. D. Thesis, Dept. of Biology, Washington Univ., St. Louis,
     Missouri, 1974.
122. ———   Electrophysiological properties of auditory neurons
     in the superior olivary complex of echolocating bats. J. comp.
     Physiol., 128, 47-56. 1978.
123. ———, Alster, H., and Suga, N.   Coordinated activity of the
     middle ear and laryngeal muscle in echolocating bats. J. Acoust.
     Soc. Am., 57, S42, VII. 1975.
124. ———, Lee, Y. H.   The just-noticeable movement speed of ob-
     stacles perceived by echolocating bats. J. Acoust. Soc. Am., 64,
     S88, HH21. 1978.
125. ———, and McCarty, J. K.   Bats avoid moving objects more
     successfully than stationary ones. Nature, London, 275, 743-744.
     1978.
126. ———, ———, and Lee, Y. H.   The avoidance of obstacles
     by little brown bats, Myotis lucifugus. Proc. Fifth Int. Bat.
     Res. Conf., (in press). 1979.
127. ———, and Ostwald, D. J.   Response of cricothyroid muscle
     to frequency-modulated sounds in FM-bats, Myotis lucifugus.
     Nature, 265, 77-78. 1977.
128. ———, ———   Further study of the electrophysiology of the
     middle-ear and cricothyroid muscles of FM bats. Nation. Sci. Coun.
     Mon. (China), V, 1014-1028. 1977.

129. Jen, P. H.-S., Ostwald, J., and Suga, N. Acoustic reflexes of middle-ear and laryngeal muscles in the FM bat Myotis lucifugus. J. Acoust. Soc. Am., 60, S4, B3. 1976.

130. ———, ———, ——— Electrophysiological properties of the acoustic middle ear and laryngeal muscle reflexes in the awake echolocating FM-bats, Myotis lucifugus. J. comp. Physiol., 124, 61-73. 1978.

131. ———, and Suga, N. Coding of directional information by neurons in the superior olivary complex of bats. J. Acoust. Soc. Am., 55, S52. 1974.

132. ———, ——— Properties of acoustic middle-ear muscle reflex in echolocating bats. J. Acoust. Soc. Am., 56, S4, B7. 1974.

133. ———, ——— Coordinated activities of middle ear and laryngeal muscles in echolocating bats. Science, 191, 950-952. 1976.

134. ———, ——— Difference between CM and N1 tuning curves in the CF-FM bats, Pteronotus parnellii rubiginosus. J. Acoust. Soc. Am., 59, S19, H13. 1976.

135. ———, ——— Functional properties of the acoustic laryngeal muscle reflex in echolocating bats. J. Acoust. Soc. Am., 59, S1, OO10. 1976.

136. ———, ——— Electrophysiological properties of middle ear and laryngeal muscle in little brown bats. Neurosci. Abst., II, 20. 1976.

137. ———, ——— Physiological properties of middle-ear and laryngeal muscles of echolocating bats. Nation. Sci. Coun. Mon. (China), IV, 2606-2617. 1976.

138. Johnson, R. A. Energy spectrum analysis as a processing mechanism for echolocation. Ph. D. Thesis, Dept. Electr. Engn., Univ. Rochester, N. Y., 1972. xv + 175 pp.

139. ———, Henson, O. W., and Goldman, L. J. Detection of insect wing beats by the bat Pteronotus parnellii. J. Acoust. Soc. Am., 55, 53. 1974.

140. ———, and Titlebaum, E. L. Energy spectrum analysis a model of echolocation processing. J. Acoust. Soc. Am. (Suppl.), 56, S39. 1974.

141. Kämper, R., and Schmidt, U. Die Morphologie der Nasenhöhle bei einigen neotropischen Chiropteren. Zoomorphologie, 87, 3-19. 1977.

142. Khajuria, H. Courtship and mating in Rhinopoma h. hardwickei Gray (Chiroptera: Rhinopomatidae). Mammalia, 36, 307-309. 1972.

143. Kindlman, P. J., Berman, L. B., Johnson, R. A., Pollak, G. D., Henson, O. W., and Novick, A. Measurement of "instantaneous" carrier frequency of bat pulses. J. Acoust. Soc. Am., 54, 1380-1382. 1973.

144. Kolb, A. Ortungsprinzip bei Fledermäusen. Z. Säugetk., 35, 306-320. 1970.

145. ———, Funktion und Wirkungsweise der Riechlaute der Mausohr-fledermaus, Myotis myotis. Z. Säugetk., 41, 226-236. 1976.

146. Kuhn, H. J. Über die Innervation des Kehlkopfes einiger Flug-hunde (Pteropodidae, Megachiroptera, Mammalia). Zool. Anz., 181, 168-181. 1968.

147. Kulzer, E., and Weigold, H. Das Verhalten der Großen Hufeisennase (Rhinolophus ferrum-equinum) bei einer Flugdressur. Z. Tierpsychol., 47, 268-280. 1978.

148. Ladik, J., and Greguss, P. Possible molecular mechanisms of
information storage in the long term memory (bat). Symp. Biol.
Hungar., 10, 343-355. 1971.

149. Long, G.R. Masked auditory thresholds from the bat, Rhinolophus ferrumequinum. J. comp. Physiol., 116, 247-255. 1977.

150. ————, and Schnitzler, H.-U. Behavioural audiograms from the
bat, Rhinolophus ferrumequinum. J. comp. Physiol., 100, 211-219.
1975.

151. Machmert, H., Theiss, D., and Schnitzler, H.-U. Konstruktion
eines Luftultraschallgebers mit konstantem Frequenzgang im Bereich
von 15 kHz bis 130 kHz. Acustica, 34, 81-85. 1975.

152. Manabe, T., Suga, N., and Ostwald, J. Aural representation in
the Doppler-shifted-CF processing area of the auditory cortex of
the mustache bat. Science, 200, 339-342. 1978.

153. Manley, G. A., Irvine, D. R. F., and Johnstone, B. M. Frequency
response of bat tympanic membrane. Nature, 237, 112-113. 1972.

154. Manley, J. A., and Johnstone, B. M. A comparison of cochlear
summating potentials in the bat and Guinea pig, including temperature effects. J. comp. Physiol., 88, 43-66. 1974.

155. McCue, J. J. G. Signal processing by the bat Myotis lucifugus.
J. Audit. Res., 9, 100-107. 1969.

156. Miline, R., Devecenski, V., and Krstic, R. Effects of auditory
stimuli on the pineal gland of hibernating bats. Acta Anat., 73
(Suppl. 56), 293-300. 1969.

157. Möhres, F. P., and Neuweiler, G. Die Ultraschallorientierung
der Großblatt-Fledermäuse (Chiroptera - Megadermatidae). Z. vergl.
Physiol., 53, 195-227. 1966.

158. Möller, J. Die Funktion von Hemmeinflüssen bei Neuronen der
lemniscalen Hörbahn bei der Echoortung von Rhinolophus ferrumequinum. Dissertation, Universität Stuttgart, 1977.

159. ———— Response characteristics of inferior colliculus neurons
of the awake CF-FM bat Rhinolophus ferrumequinum. II. Two-tone
stimulation. J. comp. Physiol. A, 125, 227-236. 1978.

160. ———— Anpassung von Neuronen der Hörbahn an die Echoortung
fliegender Hufeisennasen. Verh. Dtsch. Zool. Ges., 1978, 164,
G. Fischer Verl., Stuttgart, 1978.

161. ————, Neuweiler, G., and Zöller, H. Response characteristics
of inferior colliculus neurons of the awake CF-FM bat Rhinolophus
ferrumequinum. I. Single-tone stimulation. J. comp. Physiol. A,
125, 217-225. 1978.

162. Mogus, M. A. A theoretical approach to bat echolocation. Ph. D.
Dissertation, PA. State Univ., 1970.

163. Müller, H. C. Homing and distance-orientation in bats. Z. Tierpsychol., 23, 403-421. 1966.

164. Nakajima, Y. Fine structure of the medial nucleus of the trapezoid body of the bat with special reference to two types of synaptic
endings. J. Cell Biol., 50, 121-134. 1971.

165. Neuweiler, G.   Interaction of other sensory systems with the sonar system, in: "Animal Sonar Systems," R. G. Busnel, ed., vol. I, pp. 509-533. Lab. Physiol. Acoust., C. N. R. Z., Jouy-en-Josas, France, 1967.

166. ———— Neurophysiological investigations in the colliculus inferior of Rhinolophus ferrumequinum. Bijdr. Dierk., 40, 59-61, 1970.

167. ———— Neurophysiologische Untersuchungen zum Echoortungssystem der Großen Hufeisennase Rhinolophus ferrum equinum Schreber, 1774. Z. vergl. Physiol., 67, 273-306. 1970.

168. ———— Frequenzdiskriminierung in der Hörbahn von Säugern. Verh. Dtsch. Zool. Ges., 66, 168-176. 1973.

169. ———— Echoortung, in: "Biophysik," W. Hoppe, W. Lohmann, H. Markl, H. Ziegler, eds. Springer Verlag, Berlin, Heidelberg, New York, 1977.

170. ———— Recognition mechanisms in echolocation of bats, in: Life Sci. Res. Rep. 5, "Recognition of Complex Acoustic Signals," Th. H. Bullock, ed. Dahlem Konferenzen, Berlin, 1977.

171. ———— Die Echoortung der Fledermäuse. Rhein. Westf. Akad. Wiss., Vorträge N 272, 57-82. 1978.

172. ————, and Möhres, F. P.   The role of spacial memory in the orientation, in: "Animal Sonar Systems," R. G. Busnel, ed., vol. I, pp. 129-140. Lab. Physiol. Acoust., C.N.R.Z., Jouy-en-Josas, France, 1967.

173. ————, ———— Die Rolle des Ortsgedächtnisses bei der Orientierung der Großblatt-Fledermaus Megaderma lyra. Z. vergl. Physiol. 57, 147-171. 1967.

174. ————, Schuller, G., and Schnitzler, H.-U.   On- and off-responses in the inferior colliculus of the greater horseshoe bat to pure tones. Z. vergl. Physiol., 74, 57-63. 1971.

175. ————, and Vater, M.   Response patterns to pure tones of cochlear nucleus units in the CF-FM bat, Rhinolophus ferrumequinum. J. comp. Physiol. A, 115, 119-133. 1977.

176. Novick, A.   Echolocation of flying insects by the bat (Chilonycteris psilotis). Biol. Bull., 128, 297-314. 1965.

177. ———— Echolocation in bats: some aspects of pulse design. Amer. Sci., 59, 198-209. 1971.

178. ———— Echolocation in bats: a zoologist's view. J. Acoust. Soc. Am., 54, 139-146. 1973.

179. ———— Acoustic orientation, in: "Biology of Bats," W. A. Wimsatt, ed., vol. III. Academic Press, New York, 1977.

180. ————, and Dale, B. A.   Foraging behavior in fishing bats and their insectivorous relatives. J. Mammal., 52, 817-818. 1971.

181. ————, ———— Bats aren't all bad. National Geographic, 143, 614-637. 1973.

182. ————, and Leen, N.   "The World of Bats." Holt, Rinehart and Winston, New York, Chicago, San Francisco, 1970.

183. O'Neill, W. E., and Suga, N.   Target range-sensitive neurons in the auditory cortex of the mustache bat. Science, 203, 69-73. 1979.

184. Ostwald, J. Tonotope Organisation des Hörcortex der CF-FM-
    Fledermaus Rhinolophus ferrumequinum. Verh. Dtsch. Zool. Ges.,
    1978, 198, G. Fischer Verl., Stuttgart, 1978.
185. Patterson, A. P., and Hardin, J. W. Flight speed of five
    species of vespertilionid bats. J. Mamm., 50, 152-153. 1969.
186. Peff, T. C., and Simmons, J. A. Horizontal-angle resolution by
    echolocating bats. J. Acoust. Soc. Am., 51, 2063-2065. 1972.
187. Pirlot, P. Perspectives in sensory-ecological studies of
    higher vertebrates. Rev. Canad. Biol., 36, 337-350. 1977.
188. ————, and Pottier, J. Encephalization and quantitative brain
    composition in bats in relation to their life habits. Rev. Canad.
    Biol., 36, 321-336. 1977.
189. Pollak, G. D., Bodenhamer, R., Marsh, D. S., and Souther, A.
    Recovery cycles of single neurons in the inferior colliculus of
    unanesthetized bats obtained with frequency-modulated and constant-
    frequency sounds. J. comp. Physiol. A, 120, 215-250. 1977.
190. ————, and Henson, O. W. Specialized functional aspects of
    the middle ear muscle in the bat, Chilonycteris parnellii. J. comp.
    Physiol., 84, 167-174. 1973.
191. ————, ————, and Novick, A. Cochlear microphonic audio-
    grams in the pure tone bat Chilonycteris parnellii parnellii.
    Science, 176, 66-68. 1972.
192. ————, Marsh, D., Bodenhamer, R., and Souther, A. Echo-de-
    tecting characteristics of neurons in inferior colliculus of un-
    anesthetized bats. Science, 196, 675-678. 1977.
193. ————, ————, ————, ———— Characteristics of phasic
    "on" neurons in the inferior colliculus of unanesthetized bats
    with observations relating to mechanisms for echo-ranging. J.
    Neurophysiol., 40, 926-942. 1977.
194. ————, ————, ————, ———— A single-unit analysis of
    inferior colliculus in unanesthetized bats: Response patterns
    and spike count functions generated by CF and FM sounds. J. Neuro-
    physiol., 41, 677-691. 1978.
195. Prince, J. H. The use of echo-locating by bats. Australian
    Nat. Hist., 16, 367-370. 1970.
196. Pye, A. The structure of the cochlea in Chiroptera. I. Micro-
    chiroptera: Emballonuroidea and Rhinolophoidea. J. Morph., 118,
    495-510. 1966.
197. ———— The structure of the cochlea in Chiroptera. II. The
    Megachiroptera and Vespertilionidea of the Microchiroptera.
    J. Morph., 119, 101-120. 1966.
198. ———— The structure of the cochlea in Chiroptera. III. Micro-
    chiroptera: Phyllostomatidae. J. Morph., 121, 241-254. 1967.
199. ———— The structure of the cochlea in Chiroptera. A selection
    of Microchiroptera from Africa. J. Zool., 162, 335-343. 1970.
200. ———— The aural anatomy of bats. Bijdr. Dierk., 40, 67-70,
    1970.
201. ———— Some aspects of cochlear structure and function in bats.
    Proc. 4th Int. Bat Res. Conf., Kenya Literature Bureau, Nairobi,
    73-83. 1971.

202. Pye, A. Comparison of histological changes in the cochlea of
    the guinea pig and the bat Rousettus after acoustic trauma. Symp.
    Zool. Soc. London, 37, 97-105. Academic Press, London, 1975.
203. ———— The structure of the cochlea in Chiroptera. Periodicum
    Biologorum 75, 83-87. 1973.
204. Pye, J. D. Theories of sonar systems and their application to
    biological organisms (Discussion), in: "Animal Sonar Systems,"
    R. G. Busnel, ed., pp. 1121-1136. Lab. Physiol. Acoust., C.N.R.Z.,
    Jouy-en-Josas, France, 1967.
205. ———— Synthesizing the waveforms of bats' pulses, in: "Animal
    Sonar Systems," R. G. Busnel, ed., pp. 43-65. Lab. Physiol. Acoust.
    C. N. R. Z., Jouy-en-Josas, France, 1967.
206. ———— Bats, in: "UFAW Handbook on the Care and Maintenance of
    Laboratory Animals," 3rd edn., chap. 31, pp. 491-501. Livingstone,
    London, 1967,
207. ———— "Bats" - a natural science picture book illustrated by
    Colin Threadgall. Bodley Head, 1968.
208. ———— Animal sonar in air. Ultrasonics, 6, 32-38. 1968.
209. ———— Hearing in bats, in: "Hearing Mechanisms in Vertebrates"
    A. V. S. de Reuck and J. Knight, eds., pp. 66-84. Churchill,
    London, 1968,
210. ———— The diversity of bats. Sci. J., 5, 47-52. 1969.
211. ———— Bats and fog. Nature, 229, 572-574. 1971.
212. ———— Bimodal distribution of constant frequencies in some
    hipposiderid bats (Mammalia: Hipposideridae). J. Zool. (Lond.),
    166, 323-335. 1972.
213. ———— Echolocation by constant frequency in bats. Period.
    Biol., 75, 21-26. (Proc. 3rd Int. Bat Res. Conf., Yugoslavia, 1972)
    1973.
214. ————, and Roberts, L. H. Ear movement in a hipposiderid bat.
    Nature, 225, 285-286. 1970.
215. Radtke, S. Struktur und Verschaltung des Hörcortex der Großen
    Hufeisennase (Rhinolophus ferrumequinum). Staatsexamensarbeit,
    Fachber. Biologie, Univ. Frankfurt/M., 1979.
216. Ramprashad, F., Money, K. E., Landolt, J. P., and Laufer, J.
    A neuro-anatomical study of the cochlea of the little brown bat,
    Myotis lucifugus. J. Comp. Neurol., 178, 347-364. 1978.
217. Roberts, L. H. Variable resonance in constant frequency bats.
    J. Zool., 166, 337-348. 1972.
218. ———— Correlation of respiration and ultrasound production
    in rodents and bats. J. Zool., 168, 439-449. 1972.
219. ———— Cavity resonance in the production of orientation cries.
    Period. Biol., 75, 27-32. (Proc. 3rd Int. Bat Res. Conf.) 1973.
220. ———— Confirmation of the echolocation pulse production
    mechanisms of Rousettus. J. Mamm., 56, 218-220. 1975.
221. Roeder, K. D. Acoustic sensory responses and possible bat
    evasion tactics of certain moths. Can. Soc. Zool. Ann. Meeting
    Proc., Univ. New Brunswick, Fredericton, pp. 71-78. 1974.
222. Sales, G., and Pye, D. "Ultrasonic Communication by Animals."
    Chapman and Hall, London, 1974.

223. Schlegel, P. Directional coding by binaural brainstem units of the CF-FM bat, Rhinolophus ferrumequinum. J. comp. Physiol., 118, 327-352. 1977.

224. ———— Calibrated earphones for the echolocating bat, Rhinolophus ferrumequinum. J. comp. Physiol., 118, 353-356. 1977.

225. ————, and Suga, N. Recovery cycles of single neurons in the lateral lemniscus and inferior colliculus of echolocating bats. Psychologist, 14 (3). 1971.

226. ————, and Vater, M. Vergleichende hörphysiologische Untersuchungen an zwei echo-ortenden Fledermausarten, Molossus ater und Molossus molossus (Molossidae). Verh. Dtsch. Zool. Ges., 1978, 165, G. Fischer Verl., Stuttgart, 1978.

227. Schmidt, U., and Schmidt, Ch. Echolocation performance of the vampire bat (Desmodus rotundus). Z. Tierpsychol., 45, 349-358. 1977.

228. Schneider, R. Das Gehirn von Rousettus aegyptiacus (E. Geoffroy 1810) (Megachiroptera, Chiroptera, Mammalia). Abh. senckenb. naturf. Ges. (Frankfurt/M.), 513, 1-160. 1966.

229. Schnitzler, H.-U. Discrimination of thin wires by flying horseshoe bats (Rhinolophidae), in: "Animal Sonar Systems," R. G. Busnel, ed., pp. 69-87. Lab. Physiol. Acoust., C. N. R. Z., Jouy-en-Josas, France, 1967.

230. ———— Kompensation von Doppler-Effekten bei Hufeisen-Fledermäusen. Naturwissensch., 54, 523. 1967.

231. ———— Die Ultraschall-Ortungslaute der Hufeisen-Fledermäuse (Chiroptera-Rhinolophidae) in verschiedenen Orientierungssituationen. Z. vergl. Physiol., 57, 376-408. 1968.

232. ———— Comparison of the echolocation behavior in Rhinolophus ferrum-quinum and chilonycteris rubiginosa. Bijdr. Dierk., 40, 77-80. 1970.

233. ———— Echoortung bei der Fledermaus Chilonycteris rubiginosa. Z. vergl. Physiol., 68, 25-38. 1970.

234. ———— Fledermäuse im Windkanal. Z. vergl. Physiol., 73, 209-221. 1971.

235. ———— Die Echoortung der Fledermäuse und ihre hörphysiologischen Grundlagen. Fortschr. Zool., 21, 136-189. 1973.

236. ———— Control of Doppler shift compensation in the greater horseshoe bat, Rhinolophus ferrumequinum. J. comp. Physiol., 82. 79-92. 1973.

237. ———— Die Echoortung der Fledermäuse, in:"Grzimeks Tierleben", Ergänzungsbd.,"Verhaltensforschung", 189-204. Kindler Verlag, Zürich, 1974.

238. ———— Die Richtwirkung der Echoortung bei der Großen Hufeisennase. Verh. Dtsch. Zool. Ges. 1976, 267, G. Fischer Verl., Stuttgart, 1976.

239. ———— Die Detektion von Bewegung durch Echoortung bei Fledermäusen. Verh. Dtsch. Zool. Ges. 1978, 16-33, G. Fischer Verl., Stuttgart, 1978.

240. ————, and Grinnell, A. D. Directional sensitivity of echolocation in the horseshoe bat Rhinolophus ferrumequinum. I. Directionality of sound emission. J. comp. Physiol., 116, 51-61. 1977.

241. Schnitzler, H.-U., Schuller, G., and Neuweiler, G. Antworten des Colliculus inferior der Fledermaus Rhinolophus eureale auf tonale Reizung. Naturw., 58, 627. 1971.

242. ———, Suga, N., and Simmons, J. A. Peripheral auditory tuning for fine frequency analysis by the CF-FM bat, Rhinolophus ferrumequinum. III. Cochlear microphonics and auditory nerve responses. J. comp. Physiol. A, 106, 99-110. 1976.

243. Schuller, G. Echoortung bei Rhinolophus ferrumequinum mit frequenzmodulierten Lauten. Evoked potentials im Colliculus inferior. J. comp. Physiol., 77. 306-331. 1972.

244. ——— The role of overlap of echo with outgoing echolocation sound in the bat Rhinolophus ferrumequinum. Naturw., 61, 171-172. 1974.

245. ——— Echo delay and overlap with emitted orientation sounds and Doppler-shift compensation in the bat, Rhinolophus ferrumequinum. J. comp. Physiol. A, 114, 103-114. 1977.

246. ——— Coding of small sinusoidal frequency and amplitude modulations in the inferior colliculus of "CF-FM" bat, Rhinolophus ferrumequinum. Exp. Brain Res., 34, 117-132. 1979.

247. ———, Beuter, K., and Rübsamen, R. Dynamic properties of the compensation system of Doppler shifts in the bat Rhinolophus ferrumequinum. J. comp. Physiol., 97, 113-125. 1975.

248. ———, ———, and Schnitzler, H.-U. Response to frequency shifted artificial echoes in the bat Rhinolophus ferrumequinum. J. comp. Physiol., 89, 275-286. 1974.

249. ———, Neuweiler, G., and Schnitzler, H.-U. Collicular response to the frequency modulated final part of echolocation sounds in Rhinolophus ferrumequinum. Z. vergl. Physiol., 74, 153-155. 1971.

250. ———, and Suga, N. Storage of Doppler-shift information in the echolocation system of the "CF-FM"-bat, Rhinolophus ferrumequinum. J. comp. Physiol. A, 105, 9-14. 1976.

251. ———, ——— Laryngeal mechanisms for the emission of CF-FM sounds in the Doppler-shift compensating bat, Rhinolophus ferrumequinum. J. comp. Physiol., 107, 253-262. 1976.

252. Schweizer, H. Struktur und Verschaltung des Colliculus inferior der Grossen Hufeisennase (Rhinolophus ferrumequinum). Dissertation, Universität Frankfurt am Main, 1978.

253. Segall, W. Auditory region in bats including Icaronycteris index. Fieldiana Zool., 58, 103-108. 1971.

254. ——— The external morphology of the inner ear in bats from the phophorites of Quercy. Fieldiana Geol., 33, 59-81. 1974.

255. Shimozawa, T., Suga, N., Hendler, P., and Schuetze, S. Directional sensitivity of echolocation system in bats producing frequency modulated signals. J. Exp. Biol., 60, 53-69. 1974.

256. Shirley, D. J., and Diercks, K. J. Analysis of the frequency response of simple geometric targets. J. Acoust. Soc. Am., 48, 1275-1282. 1971.

257. Simmons, J. A. The sonar sight of bats. Psychol. Today, 2, 50-57. 1968.

258. Simmons, J. A. Depth perception by sonar in the bat Eptesicus fuscus. Ph. D. Dissertation, Princeton Univ., (x + 208 pp.), 1968.

259. ──────── Acoustic radiation patterns for the echolocating bats, Chilonycteris rubiginosa and Eptesicus fuscus. J. Acoust. Soc. Am., 46, 1054-1056. 1969.

260. ──────── The sonar receiver of the bat. Ann. N. Y. Acad. Sci., 188, 161-174. 1971.

261. ──────── Narrow band CF-FM echolocation in bats (Summary). J. Acoust. Soc. Am., 50, 148. 1971.

262. ──────── Echolocation in bats: Signal processing of echoes for target range. Science, 171, 925-928. 1971.

263. ──────── The resolution of target range by echolocating bats. J. Acoust. Soc. Am., 54, 157-173. 1973.

264. ──────── Response of the Doppler echolocation system in the bat, Rhinolophus ferrumequinum. J. Acoust. Soc. Am., 56, 672-682. 1974.

265. ──────── (Rapporteur), et al. Localization and identification of acoustic signals, with reference to echolocation. Group report in: Life Sci. Res. Rep. 5, "Recognition of Complex Acoustic Signals," Th. H. Bullock, ed. Dahlem Konferenzen, Berlin, 1977.

266. ────────, Fenton, M. B., and O'Farrell, M. J. Echolocation and pursuit of prey by bats. Science, 203, 16-21. 1979.

267. ────────, Howell, D. J., and Suga, N. Information content of bat sonar echoes. Amer. Sci., 63, 204-215. 1975.

268. ────────, Lavender, W. A., Lavender, B. A., et al. Target structure and echo spectral discrimination by echolocating bats. Science, 186, 1130-1132. 1974.

269. ────────, ────────, ────────, et al. Echolocation by free-tailed bats (Tadarida). J. comp. Physiol., 125, 291-299. 1978.

270. ────────, and O'Farrell, M. J. Echolocation by the long eared bat, Plecotus phyllotis. J. comp. Physiol., 122, 201-214. 1977.

271. ────────, and Vernon, J. A. Echolocation discrimination of targets by the bat Eptesicus fuscus. J. Exp. Zool., 176, 315-328. 1971.

272. Skinner, D. P., Altes, R. A., and Jones, J. D. Broadband target classification using a bionic sonar. J. Acoust. Soc. Am., 62, 1239-1246, 1977.

273. Small, R., and Levine, R. R. Interspecific communication between bats and moth, in: "Course in Animal Communication," R. G. Busnel, ed. The City College, Paris, 1971.

274. Smith, D. M., Mercer, R. M., Goldman, L. H., Henson, W. W., and Henson, M. M. Phase-locked loop device for the fine frequency analysis of the biosonar signals of bats. J. Acoust. Soc. Am., 61, 1092-1093. 1977,

275. Stewart, J. L., and Kasson, J. M. Simulation mechanisms in animal echo ranging. U. S. Gov. Research Dev. Rep., 70. 1970.

276. Stones, R. C., and Branick, L. P. Use of hearing in homing by two species of Myotis bats. J. Mamm., 50, 157-160. 1969.

277. Strother, G. K. Comments on aural pulse compression in bats and humans. J. Acoust. Soc. Amer., 41, 529. 1967.

278. Strother, G. K., and Mogus, M. Acoustical beam patterns for bats: Some theoretical considerations. J. Acoust. Soc. Am., 48, 1430-1432. 1970.

279. Suga, N. Analysis of frequency-modulated sounds by auditory neurons of echolocating bats. J. Physiol., 179, 26-53. 1965.

280. ———— Functional properties of auditory neurons in the cortex of echolocating bats. J. Physiol., 181, 671-700. 1965.

281. ———— Responses of cortical auditory neurons to frequency modulated sounds in echo-locating bats. Nature, 206, 890-891. 1965.

282. ———— Analysis of frequency-modulated and complex sounds by single auditory neurons of bats. J. Physiol., 198, 51-80. 1968.

283. ———— Classification of inferior collicular neurons of bats in terms of responses to pure tones, FM sounds, and noise bursts. J. Physiol., 200, 555-574. 1969.

284. ———— Echo-location and evoked potentials of bats after ablation of inferior colliculus. J. Physiol., 203, 707-728. 1969.

285. ———— Echo-location of bats after ablation of auditory cortex. J. Physiol., 203, 729-739. 1969.

286. ———— Echo-ranging neurons in the inferior colliculus of bats. Science, 170, 449-451. 1970.

287. ———— Responses of inferior collicular neurons of bats to tone bursts with different rise times. J. Physiol., 217, 159-177. 1971.

288. ———— Analysis of information-bearing elements in complex sounds by auditory neurons of bats. Audiol., 11, 58-72, 1972.

289. ———— Neurophysiological analysis of echolocation in bats, in: "Animal Orientation and Navigation," S. R. Galler et al., eds. NASA Press, Washington, D. C., 1972.

290. ———— Feature extraction in the auditory system of bats, in: "Basic Mechanisms in Hearing," A. R. Møller, ed., pp. 675-744. Academic Press, New York and London, 1973.

291. ———— Amplitude spectrum representation in the Doppler-shifted-CF processing area of the auditory cortex of the mustache bat. Science, 196, 64-67. 1977.

292. ———— Specialization of the auditory system for reception and processing of species-specific sounds. Federation Proc., 37, 2342-2354. 1978.

293. ————, and Jen, P. H.-S. Muscular control of incoming signals in the auditory system of echolocating bats. J. Acoust. Soc. Am., 55, S52, AA9. 1974.

294. ————, ———— Peripheral control of acoustic signals in the auditory system of echolocating bats. J. Exp. Biol., 62, 277-311. 1975.

295. ————, ———— Disproportionate tonotopic representation for processing CF-FM sonar signals in the mustache bat auditory cortex. Science, 194, 542-544. 1976.

296. ————, ———— Neural vs. mechanical tuning curves in the CF-FM bats Pteronotus parnellii rubiginosus. J. Acoust. Soc. Am., 59, Suppl. No. 1, S18. 1976.

297. Suga, N., and Jen, P. H.-S. Further studies on the peripheral auditory system of CF-FM bats specialized for fine frequency analysis of Doppler-shifted echoes. J. Exp. Biol., 69, 207-232. 1977.

298. ————, Neuweiler, G., and Möller, J. Peripheral auditory tuning for fine frequency analysis by the CF-FM bat, Rhinolophus ferrumequinum. IV. Properties of peripheral auditory neurons. J. comp. Physiol. A, 106, 111-125. 1976.

299. ————, ————, Jen, P. H.-S., and Möller, J. Adaptation of the peripheral auditory system of CF-FM bats for reception and analysis of predominant components in orientation sounds and echoes. J. Acoust. Soc. Am., 59, Suppl. No. 1, S46. 1976.

300. ————, and O'Neill, W. E. Mechanisms of echolocation in bats - comments on the neuroethology of the biosonar system of CF-FM bats. Trends Neurosc., 1, 35-38. 1978.

301. ————, ————, and Manabe, T. Cortical neurons sensitive to combinations of information-bearing elements of biosonar signals in the mustache bat. Science, 200, 778-781. 1978.

302. ————, ————, ———— Harmonic-sensitive neurons in the auditory cortex of the mustache bat. Science, 203, 270-274. 1979.

303. ————, and Schlegel, P. Neural attenuation of responses to emitted sounds in echolocating bats. Science, 177, 82-84. 1972.

304. ————, ———— Coding and processing in the auditory systems of FM-Signal producing bats. J. Acoust. Soc. Am., 54, 174-190. 1973.

305. ————, ————, Shimozawa, T., and Simmons, J. A. Orientation sounds evoked from echolocating bats by electrical stimulation of the brain. J. Acoust. Soc. Am., 54, 793-797. 1973.

306. ————, Schuller, G., and Jen, P. H.-S. Neurophysiological studies on the echolocation system of bats sensitive to Doppler-shifted echoes. J. Acoust. Soc. Am., 60, S4, B2. 1976.

307. ————, and Shimozawa, T. Site of neural attenuation of responses to self-vocalized sounds in echolocating bats. Science, 183, 1211-1213. 1974.

308. ————, Simmons, J. A., and Jen, P. H.-S. Peripheral specialization for fine analysis of Doppler-shifted echoes in the auditory system of the "CF-FM" bat Pteronotus parnellii. J. Exp. Biol., 63, 161-192. 1975.

309. ————, ————, ———— Peripheral specialization for analysis of Doppler-shifted echoes in the auditory system of the "CF-FM" bat, Pteronotus parnellii. I. Cochlear microphonic. J. Acoust. Soc. Am., 57, S42. 1975.

310. ————, ————, ———— Peripheral specialization for analysis of Doppler-shifted echoes in the auditory system of the "CF-FM" bat, Pteronotus parnellii. II. Properties of peripheral auditory neurons. J. Acoust. Soc. Am., 57, S42. 1975.

311. ————, ————, Shimozawa, T. Neurophysiological studies on echolocation systems in awake bats producing CF-FM orientation sounds. J. Exp. Biol., 61, 379-399. 1974.

312. Suthers, R. A. Acoustic orientation by fish-catching bats. J. Exp. Zool, 158, 319-348. 1965.

313. ———— Comparative echolocation by fishing bats. J. Mamm., 48, 79-87. 1967.

314. ———— Wie orten fischfangende Fledermäuse ihre Beute? Umschau i. W. u. T., 67, 693-696. 1967.

315. ———— A comment on the role of choroidal papillae in the fruit bat retina. Vision Res., 10, 921-922. 1970.

316. ———— Vision, olfaction, taste, in: "Biology of Bats," W. Wimsatt, Ed., vol. II. Academic Press, New York, 1970.

317. ———— Responses of neurons in the visual cortex of the echo-locating bat, Rousettus aegypticus. Proc. East Afr. Acad.

318. ————, and Chase, J. Visual pattern discrimination by an echolocating bat. Amer. Zool., 6, 573. 1966.

319. ————, ————, and Braford, B. Visual form discrimination by echolocating bats. Biol. Bull, 137, 535-546. 1969.

320. ————, and Fattu, J. M. Fishing behaviour and acoustic orientation by the bat (Noctilio labialis). Anim. Behav., 21, 61-66. 1973.

321. ————, ———— Mechanisms of sound production by echolocating bats. Amer. Zool., 13, 1215-1226. 1973.

322. ————, Thomas, St. P., and Suthers, B. J. Respiration, wingbeat and ultrasonic pulse emission in an echo-locating bat. J. Exp. Biol., 56, 37-48. 1972.

323. ————, and Wallis, N. E. Optics of the eye of some echolocating bats. Vision Res., 10, 1165-1173. 1970.

324. Tepaske, E. R. Some morphological aspects and taxonomic relationship of the middle ear in bats. Diss. Abstr., 26, 65-8784 (155 pp.) 1966.

325. Vater, M., and Schlegel, P. Antwortcharakteristika einzelner Neurone im Colliculus inferior der echoortenden Fledermausarten Molossus ater und Molossus molossus auf tonale und FM Signale. Verh. Dtsch. Zool. Ges., 1978, 197, G. Fischer Verl., Stuttgart, 1978.

326. Vernon, J. A., Dalland, J. S., and Wever, E. G. Further studies of hearing in the bat Myotis lucifugus, by means of cochlear potentials. J. Aud. Res., 6, 153-163. 1966.

327. ————, and Peterson, E. A. Echolocation signals in the free-tailed bat, Tadarida mexicana. J. Aud. Res., 5, 317-330. 1965.

328. ————, ———— Hearing in the vampire bat, Desmodus rotundus, as shown by cochlear potentials. J. Aud. Res., 6, 181-187. 1966.

329. Wassif, K., and Madkour, G. M. The anatomy of the hyoid bone, larynx, and upper part of the trachea in some Egyptian bats. Bull. Zool. Soc. Egypt., 22, 15-26. 1969.

330. Webster, F. A. Some acoustical differences between bats and men. Int. Conf. on Sens. Devices for the Blind, St. Dunstan's, London, 63-87. 1967.

331. Webster, F. A., and Brazier, O. G. Experimental studies on target detection, evaluation, and interception by echolocating bats. T. D. R. No. AMRL-TR-65-172, Aerospace Medical Div., U. S. A. F. Systems Command. 1965.

332. ————, ———— Experimental studies on echolocation mechanisms in bats. T. D. R. No. AMRL-TR-67-192, Aerospace Medical Div., U. S. A. F. Systems Command. 1968.

333. ————, ———— Echolocation investigations on bats and humans: Target localization evaluation. Sensory Sys. Lab., Tucson Ariz., AD-697 070, pp. 80. 1968.

334. Wickler, W., and Seibt, U. Doppelklick-Orientierungslaute bei einem Epauletten-Flughund. Naturw., 61, 367. 1974.

335. Williams, T. C. Nocturnal orientation techniques of a neotropical bat. Ph. D. Thesis, Rockefeller Univ., New York, 1968.

336. ————, and Williams, J. M. Radio tracking of homing bats. Science, 155, 1435-1436. 1967.

337. ————, ———— Radio tracking of homing and feeding flights of a neotropical bat Phyllostomus hastatus. Anim. Behav., 18, 302-309. 1970.

338. ————, ————, and Griffin, D. R. The homing ability of the neotropical bat Phyllostomus hastatus with evidence for visual orientation. Anim. Behav., 14, 468-473. 1966.

339. Wilson, J. P. Towards a model for cochlear frequency analysis, in: "Psychophysics and Physiol. Hearing," E. F. Evans and J. P. Wilson, eds. Academic Press, New York, 1977.

340. Wimsatt, W. A., ed. "Biology of Bats," vol. I - III. Academic Press, New York, 1970.

341. Woolf, N. The ontogeny of bat sonar sounds: With special emphasis on sensory deprivation. Sc. D. Thesis, J. Hopkins Univ., Baltimore, 1974.

342. Yalden, D. W., and Morris, P. A. "The Lives of Bats." Quadrangle/The New York Times Book Co., New York, 1975.

BIBLIOGRAPHY OF ECHOLOCATION PAPERS ON AQUATIC MAMMALS PUBLISHED
BETWEEN 1966 AND 1978

Abramov, A. P., Golubkov, A. G., Yershova, I. V., Fradkin, V. G.
and Korolev, V. I., 1972, Investigation of the dolphin's ability to differentiate the volume of objects according to
dimensions and materials, in: "Problemy Biologicheskey
Kibernetiki", A. I. Berg, ed., Isdatelstvo Nauka, Leningrad.

Abramov, A. P., Ayrapet'yants, E. Sh., Burdin, V. I., Golubkov,
A. G., Yershova, I. V., Zhezherin, A. R., Korolev, V. I.,
Malyshev, Yu. A., Ul'yanov, G. K., and Fradkin, V. G., 1971,
Investigations of delphinid capacity to differentiate between
three-dimensional objects according to linear size and
material, "Report from the 7th All-Union Acoustical Confer-
ence", Leningrad, as cited in: Bel'kovich, V. M. and Dubrovskiy,
N.A., 1976, "Sensory Bases of Cetacean Orientation", Nauka,
Leningrad (English translation JPRS L/7157).

Ackman, R. G., Eaton, C. A., and Litchfield, C., 1971, Composition
of wax esters, triglycerides and diacyl glyceryl ethers
in the jaw and blubber fats of the Amazon river dolphin
(Inia geoffrensis), Lipids, 6:69.

Ackman, R. G., Sipos, J. C., Litchfield, C., and Hilaman, B.,
1972, Characterization of unusual wax esters from the jaw
fat of the Atlantic bottle-nosed dolphin (Tursiops truncatus),
Jour. Am. Oil Chem. Soc., 49:305.

Ackman, R. G., Sipos, J. C., Eaton, C. A., Hilaman, B. L. and
Litchfield, C., 1973, Molecular species of wax esters within
jaw fat of the Atlantic bottlenosed dolphin, Tursiops
truncatus, Lipids, 8:661.

Agarkov, G. B., 1969, Soviet cetacean morphology research outlined,
Vestnik Akademii Nauk, 38(8):58.

Agarkov, G. B. and Khomenko, B. G., 1973, Morphological and functional analysis of certain sensory areas of the dolphin's
head, IV-oe Vsesoyuznoye soveshchaniye po Bionike, 4:1.

Agarkov, G. B. and Valiulina, F. G., 1974, The issue of the innervation of the mandibular region of the common dolphin,
Bionika, 8:117.

Agarkov, G. B., Khomenko, B. G., and Khadzhinskiy, V. G., 1974,
in: "Morphology of Delphinidae", Izdatel'stvo Naukova
Dumka, Kiev.

Agarkov, G. B., Solukha, B. V., and Khomenko, B. G., 1971, Echolocation capacity of dolphins, Bionika, 5:52.

Agarkov, G. B., Solukha, B. V., and Khomenko, B. G., 1973, On the
hydrolocation capability of dolphins, National Translation
Center, 75:7.

Agarkov, G. B., Solukha, B. V., and Zakletskiy, A. V., 1975, Convergence in structure and function of the auditory system of
fish, amphibians, reptiles, birds and mammals, in: "Morskiye
Mlekopitayushchiye", G. B. Agarkov, ed., Izdatel'stvo
Naukova Dumka, Kiev.

Alexeyeva, T. V., Golubkov, A. G., and Ershova, I. V., 1971, On the problem of the active width of the spectrum of dolphins' echolocation signals, Trudy Akusticheskogo Instituta 17:99.

Altes, R. A., 1971, Computer derivation of some dolphin echolocation signals, Science, 173:912.

Altes, R. A., 1973, Study of animal signals and neural processing with application to advanced sonar systems, ESL Inc. Rpt. No. 115.

Altes, R. A., 1974, Study of Animal Sensory Systems with Application to Sonar, ESL Inc. Rpt. No. 144.

Altes, R. A., 1976, Sonar for generalized target description and its similarity to animal echolocation systems, Jour. Acous. Soc. Amer., 59:97.

Altes, R. A., 1977, Localization in cetacean echolocation, in: "Proceedings (Abstracts) Second Conference On The Biology of Marine Mammals", San Diego.

Altes, R. A., and Skinner, D. P., 1977, Sonar velocity resolution with a linear-period-modulated pulse, Jour. Acoust. Soc. Amer., 61:1019.

Altes, R. A., Evans, W. E., and Johnson, C. S., 1975, Cetacean echolocation signals and a new model for the human glottal pulse, Jour. Acoust. Soc. Amer., 57:1221.

Andersen, H. T. (ed.), 1969, "The Biology of Marine Mammals", Academic Press, New York.

Andersen, S., 1970, Auditory sensitivity of the harbour porpoise Phocoena phocoena, in: "Investigations on Cetacea", G. Pilleri, ed., Bentelli AG., Berne.

Andersen, S., 1970, Directional hearing in the harbour porpoise Phocoena phocoena, in: "Investigations on Cetacea Vol. 2", G. Pilleri, ed., Bentelli AG., Berne.

Andersen, S., 1971, Orientierung hos delphiner, Nat. Verden, 5:181.

Andersen, S. and Pilleri, G., 1970, Audible sound production in captive Platanista gangetica., in: "Investigations on Cetacea Vol. 2", G. Pelleri, ed., Bentelli AG., Berne.

Asa-Dorian, P. V. and Perkins, P. J., 1967, The controversial production of sound by the California gray whale, Eschrichtius robustus Norsk Hvalfangsttid, 56:74.

Au, W. W. L., 1977, Target analysis using simulated dolphin echolocation signals, in: "Proceedings (Abstracts) Second Conference on the Biology of Marine Mammals, San Diego.

Au, W. W. L. and Hammer, C., 1978, Analysis of target recognition via echolocation by an Atlantic bottlenose porpoise (Tursiops truncatus), Jour. Acoust. Soc. Amer., 60:587.

Au, W. W. L., Floyd, R. W., and Haun, J. E., 1978, Propagation of Atlantic bottlenose dolphin echolocation signals, Jour. Acoust. Soc. Amer., 64:411.

Au, W. W. L., Floyd, R. W., Penner, R. H. and Murchison, A. E.,
    1974, Measurement of echolocation signals of the Atlantic
    bottle-nose dolphin, Tursiops truncatus (Montagu), in open
    waters, Jour. Acoust. Soc. Amer., 56:1280.

Audouin, F., 1967, Etude bionique de l'auto-information par
    echoacoustiques dans le regne animal, Rapport techn.du
    Marche DRME n°66-34179-00-480-75-05, Thomson-Houston, Vol. II.

Awbrey, F. T., 1979, Background study of acoustical and
    bioacoustical factors in tuna fishing productivity and
    associated porpoise mortality, Fish. Bull. (in press).

Ayrapet'yants, E. Sh. and Konstantinov, A. I., 1970, "Echolocation
    in Nature", Nauka, Leningrad (English translation JPRS
    63328).

Ayrapet'yants, E. Sh. and Konstantinov, A. I., 1976, Physiological
    investigations of ultrasonic echolocation in animals, paper
    presented at the 25th International Congress of Physiol.
    Sciences, Munich.

Ayrapet'yants, E. Sh., Konstantinov, A. I., and Matjushkin, D. P.,
    1969, Brain echolocation mechanisms and bionics, Acta Physiol.
    Acad. Sci. Hung., 35:1.

Ayrapet'yants, E. Sh., Golubkov, A. G., Yershova, I. V.,
    Zhezherin, A. R., Zvorkin, V. N., and Korolev, V. I., 1969,
    Echolocation differentiation and characteristics of radiated
    pulses in dolphins, Report of the Academy of Science of
    the USSR, 188:1197 (English translation JPRS 49479).

Ayrapet'yants, E. Sh., Voronov, V. A., Ivanenko, Y. V., Ivanov,
    M. P., Ordovskii, D. L., Sergeev, B. F., and Chilingiris, V.
    I., 1973, The physiology of the sonar system of Black Sea
    dolphins, Jour. Evol. Biochem. Physiol., 9:364 (English
    translation JPRS 60298).

Babkin, V. P., and Dubrovskiy, N.A., 1971, Range of action
    and noise stability of the echolocation system of the bottle-
    nose dolphin in detection of various targets, Tr. Akust.
    Inst., Moscow, 17:29, as cited in: Bel'kovich, V. M. and
    Dubrovskiy, N. A., 1976, "Sensory Bases of Cetacean Orienta-
    tion", Nauka, Leningrad (English translation JPRS L/7157).

Babkin, V. P., Dubrovskiy, N. A., Krasnov, P. S, and Titov, A.
    A., 1971, Discrimination of material of spherical targets by
    the bottlenose Dolphin, in: "Report from the 7th All-Union
    Acoustical Conference, Leningrad".

Backus, R. H. and Schevill, W. E., 1966, Physeter clicks, in:
    "Whales, Dolphins, and Porpoises", K. S. Norris, ed.,
    Univ. California Press, Berkeley.

Bagdonas, A., Bel'kovich, V. M., and Krushinskaya, N. L., 1970,
    Interaction of analyzers in dolphins during discrimination
    of geometrical figures under water, Jour. Higher Neural
    Act., 20:1070 (English translation in: "A Collection of
    Translations of Foreign Language Papers on the Subject of
    Biological Sonar Systems", K. J. Diercks, 1974, ed., Applied
    Research Lab, U. of Texas, Austin, Tech. Rept. 74-9).

Barham, E. G., 1973, Whales' respiratory volume as a possible
    resonant receiver for 20 Hz signals, Nature, 245:220.
Barta, R. E., 1969, Acoustical pattern discrimination by an
    Atlantic bottlenosed dolphin, unpublished manuscript, Naval
    Undersea Center, San Diego, CA.
Bastian, J., 1967, The transmission of arbitrary environmental
    information between bottle-nose dolphins, U. S. Naval Ordnance
    Test Stat., China Lake, Calif., NOTS TP 4117.
Bastian, J., Wall, C., and Anderson, C. L., 1968, Further investiga-
    tions of the transmission of arbitrary information between
    bottlenose dolphins, NUWC TP 109.
Bateson, G., 1966, Problems in cetacean and other mammalian
    communication, in: "Whales, Dolphins, and Porpoises", K. S.
    Norris, ed., Univ. of Calif., Berkeley.
Beach, F. A. III, and Herman, L. M., 1972, Preliminary studies of
    auditory problem solving and intertask transfer by the
    bottlenose dolphin, Psychological Record, 22:49.
Beach, F. A. III, Pepper, R. L., Nachtigall, P. E., Simmons, J. V.,
    and Siri, P. A., 1974, Spatial habit reversal in two species
    of marine mammals, Psychological Record, 24:385.
Beamish, P., 1971, Biological sonar of diving mammals, an analyti-
    cal model, in: "Proc. 7th Annu. Conf. Biol. Sonar and Diving
    Mamm., Stanf. Res. Inst.", Menlo Park, Calif.
Beamish, P., 1972, Whale sonar, in: "Proc. 10th Annu. Meeting
    Canadian Committee on Acoustics", R. J. Donato, ed.,
    Nation. Res. Council, Ottawa.
Beamish, P., and Mitchell, E., 1971, Ultrasonic sounds recorded
    in the presence of a blue whale Balaenoptera musculus, Deep-
    Sea Res., 18:803.
Beamish, P., and Mitchell, E., 1973, Short pulse length audio
    frequency sounds recorded in the presence of a Minke whale
    Balaenoptera acutorostrate, Deep-Sea Res., 20:375.
Bel'kovich, V. M., 1970, About "Acoustical Vision" in the dolphin.
    Experimental matters. Ekologia, 6:89.
Bel'kovich, V. M. and Borisov, V. I., 1971, Locational discrimi-
    nation of figures of complex configuration by dolphins,
    Trudy Akusticheskogo Institute, 17:19.
Bel'kovich, V. M. and Dubrovskiy, N. A.,1976, "Sensory Bases of
    Cetacean Orientation", Nauka, Leningrad (English translation
    JPRS L/7157).
Bel'kovich, V. M. and Gurevich, V. S. 1969, The whales (Cetacea,
    Odontoceti) are the experimental animals, "Scientific
    Conference on Problems of the World Ocean", M:16.
Bel'kovich, V. M. and Gurevich, V. S., 1970, Problems on keeping
    of the dolphins in captivity and performing experimental
    work with them, in: "Collection Material of the Conference
    'Ocean'", Moscow:108-112.
Bel'kovich, V. M. and Nesterenko, I., 1971, How the dolphin's
    locator operates, Priroda., 7:71 (German translation by W.
    Petri, 1974, Naturwissenschaftliche Rundschau, 0004:0143).

Bel'kovich, V. M., and Resnikov, A. Ye., 1971, What's new in dolphin sonar, Priroda, 11:84.

Bel'kovich, V. M. and Solntzeva, G. N., 1970, Anatomy and function of the ear in the dolphin, Zoolischeskiy Zhurnal, 49(2):275. (English translation JPRS 50253).

Bel'kovich, V. M., Borisov, V. I., Gurevich, V. S., 1970, Angular resolution by echolocation in Delphinus delphis, in: "Proceedings of Scientific Technical Conference, Ministry of Higher and Secondary Specialized Education RSFSR, Leningrad" (English translation in: "A Collection of Translations of Foreign Language Papers on the Subject of Biological Sonar Systems", K. J. Diercks, 1974, ed., Applied Research Lab, Univ. of Texas, Austin, Tech. Rept. 74-9).

Bel'kovich, V. M., Gurevich, V. S., and Borisov, V. I., 1970, Discrimination by the dolphins of the geometrical figures according to the angle, in: "The 23rd Scientific-Technical Conference of the LIAP", Leningrad.

Bel'kovich, V. M., Andreyev, F. V., Vronskaya, S. D., and Cherdantsev, A. I., 1975, Investigation of orientation and behavior of wild bottlenose dolphins (Tursiops truncatus), in: "Morskiye Mlekopitayushchiye", G. B. Agarkov, ed., Isdatel'stvo Naukova Dumka Kiev.

Bel'kovich, V. M., Bagdonas, A. P., Gurevich, V. S., and Krushinskaya, N. L., 1969, Echolocation of the toothed whales (Cetacea), "Scientific Conference of Problems of the World Ocean".

Bel'kovich, V. M., Borisov, V. I., Gurevich, V. S., and Krushinskaya, N. L., 1969, Echolocating capabilities of the common dolphin (Delphinus delphis), Zoologicheskiy Zhurnal, 48:876 (English translation JPRS 48780).

Bel'kovich, V. M., Borisov, V. I., Gurevich, V. S., Krushinskaya, N. L., and Dmitrieva, I. L., 1972, Echolocation discrimination of distance between the targets (angular resolution) by dolphins, in: "Proceedings of the 5th All-Union Conference on Marine Mammals", Makhachkala, Part 2:30.

Bel'kovich, V. M., Gurevich, V. S., Dmitrieva, I. L., Borisov, V. I., and Krushinskaya, N. L., 1973, Echolocation capabilities of the dolphins discriminating the geometrical figures by the square, in: "Material of the 6th All-Union Conference on Studying of Marine Mammals, Makhachkala: 57.

Berzin, A. A., 1971, "The Sperm Whale". Pacific Scientific Research Institute of Fisheries and Oceanography (USSR), (Trans. by Israel Program for Scientific Translations, publ. by U. S. National Technical Information Service, Springfield, Va.).

Blevins, C. E., and Parkins, B., 1973, Functional anatomy of the porpoise larynx, Amer. J. Anat., 138:151.

Blomberg, J., 1972, Pilot whale head oil: lipid analysis and ultrasonic studies, in: "Program of the XIth Congr. Int. Soc. for Fat Res.", Goteborg, Abstract 223.

Bloome, K. A., 1969, An electron microscopic study of the organ
    of corti of the porpoise (Abstract), Am. Zool., 9:1148.
Bogoslovskaya, L. S., 1974, Distinctions of neuronal structure
    of primary acoustic centers in the dolphins, in: "Morfologiya
    Fiziologiya i Akustika Morskikh Mlekopitayushchikh",
    V. Ye. Sokolov, ed., Izdatel'stvo Nauka, Moscow.
Bogoslovskaya, L. S., 1975, Ultrastructure of the spiral ganglion
    and acoustic nerve in bottlenose dolphins (Tursiops truncatus),
    in: "Morskiye Mlekopitayushchiye", G. B. Agarkov, ed.,
    Izdatl'stvo Naukova Dumka, Kiev.
Bourcart, 1970, le dauphin, Rev. des Corps de Sante, 11:311.
Bounton, K. L., 1966, The remarkable dolphin (ear, sonar, echos),
    III: Sound senders and receivers, Can. Audubon., 28:24.
Brown, A. M. and Pye, J. D., 1975, Auditory sensitivity at high
    frequencies in mammals, in: "Advances in Comparative Physiol-
    ogy and Biochemistry, Vol. 6", O. Lowenstein, ed., Academic
    Press, New York.
Bullock, T. H. and Gurevich, V. S., 1979, Soviet Literature on
    the nervous system and psychobiology of cetaceans, Interna-
    tional Review of Neurobiology, (In Press).
Bullock, T. H. and Ridgway, S. H., 1972, Evoked potentials in the
    central auditory system of alert porpoises to their own and
    artificial sounds, J. Neurobiol., 3:79.
Bullock, T. H. and Ridgway, S. H., 1972, Neurophysiological
    findings relevant to echolocation in marine mammals, in:
    "Animal Orientation and Navigation, S. R. Galles, K.,
    Schmidt-Konig, G. J. Jacobs, and R. E. Belleville, eds., NASA,
    Washington, D. C.
Bullock, T. H., Ridgway, S. H., and Suga, N., 1971, Acoustically
    evoked potentials in midbrain auditory structures in sea
    lions (Pinnipedia), Z. vergl. Physiol., 74:732.
Bullock, T. H., Grinnell, A. D., Ikezono, E., Kameda, K.,
    Katsuki, Y., Nomoto, M., Sato, O., Suga, N., and Yanagisawa, K.,
    1968, Electrophysiological studies of central auditory
    mechanisms in cetaceans, Zeitschrift fur vergieichende
    Physiologie, 59:117.
Burdin, V. I., Reznik, A. M., Skornyakov, V. M. and Chupakov, A. G.
    1974, Study of communicative signals in Black Sea dolphins,
    Akusticheskiy Zhurnal, 20(4):518.
Burdin, V. I., Reznik, A. M., Skornyakov, V. M. and Chupakov,
    A. G., 1975, Communication signals of the black sea bottle-
    nose dolphin, Sov. Phys. Acoust., 20:314.
Burdin, V. I., Markov, V. I., Reznik, A. M., Skornyakov, V. M. and
    Chupakov, A. G., 1971, Ability of Tursiops truncatus (Ponticus
    barabash) to distinguish a useful signal against a noise back-
    ground, in: "Morfologiya i Ekologiya Morskikh
    Mlekopitayushchikh", V. Ye. Sokolov, ed., Izdatl'stvo
    Nauka, Moscow.

Burdin, V. I., Markov, V. I., Reznik, A. M., Skornyakov, V. M., and Chupakov, A. G., 1971, Determination of the just noticeable intensity difference for white noise in the Black Sea dolphin, in: "Morfologiya i Ekologiya Morskikh Mlekopitayushchikh", V. Ye. Sokolov, ed., Izdatl'stvo Nauka Moscow.

Burdin, V. I., Markov, V. I., Reznik, A. M., Skornyakov, V. M., and Chupakov, A. G., 1971, Determination of the just noticeable intensity difference for white noise in the bottlenose dolphin (Tursiops truncatus Barabash), in: "Morphology and Ecology of Marine Mammals", K. K. Chapskii and V. Y. Sokolov, eds., Wiley, New York.

Busnel, R.-G., 1966, Information in the human whistled language and sea mammal whistling, in: "Whales, Dolphins and Porpoises", K. S. Norris, ed., Univ. Calif. Press, Berkeley.

Busnel, R.-G., 1970, Themes de reflexions a propos de l'echolocation biologique et de la bionique, in: "Principles and Practice of Bionics", H. E. Von Gierke, W. D. Keidel, H. L. Oestreicher, eds., Technivision Serv., Slough - G. B.

Busnel, R.-G., 1973, Symbiotic relationship between man and dolphins Trans. N. Y. Acad. Sci., Series II, 25:112.

Busnel, R.-G., and Dziedzic, A., 1966, Acoustic physiologique - caracteristiques physiques de certains signaux acoustiques due Delphinide Steno bredansis LESSON, C. R. Acad. Sci., 262:143.

Busnel, R.-G., Dziedzic, A., 1966, Acoustic signals of the pilot whale Glogicephala melaena and the porpoises Delphinus delphis and Phocoena phocoena, in: "Whales, Dolphins and Porpoises", K. S. Norris, ed., U. of Calif. Press, Berkeley.

Busnel, R.-G. and Dziedzic, A., 1967, Observations sur le comportement et les emissions acoustiques du cachalot lors de la chasse, Bocagiana, Mus. Municipal do Funchal, Madeira, 14:1.

Busnel, R.-G. and Dziedzic, A., 1967, Resultats metrologigues experimentaux de l'echolocation chez le Phocaena phocaena at leur comparison avec ceux de certaines chauves - souris, in: "Animal Sonar Systems, Biology and Bionics", R.-G. Busnel, ed., Laboratoire de Physiologie Acoustique, Jouy-en-Josas, France.

Busnel, R.-G. and Dziedzic, A., 1968, Caracteristiques physiques des signaux acoustiques de Pseudorca crassidens OWEN (Cetace odontocete), Extr. d. Mammal, 32:1.

Busnel, R.-G., and Dziedzic, A., 1968, Etude des signaux acoustiques associes a des situations de detresse chez certains cetaces odontocetes, Ann. de L'Inst. Oceanographique, Nouvelle Serie, 46:109.

Busnel, R.-G., Dziedzic, A., and Alcuri, A., 1974, Etudes preliminaires de signaux acoustiques du Pontoporia blainvillei CERVAIS et d' ORGIBNY (Cetacea, Platanistidae), Mammalia, 38:449.

Busnel, R.-G, Dziedzic, A., and Escudie, B., 1969, Autocorrelation
    et analyse spectrale des signauz "sonar" de deux especes
    Cetaces Odontocetes utilisant les basses frequences,
    C. R. Acad. Sci. Paris, 269:365.
Busnel, R.-G., Pilleri, G., and Fraser, F. C., 1968, Notes
    concerant le dauphin Stenella styx Gray 1846, Extrait de
    Mammalia, 32:192.
Busnel, R.-G., Escudie, B., Dziedzic, A. and Hellion, A., 1971,
    Structure des clics doubles d'echolocation du Globicephale
    (Cetace Odontocete), C. R. Acad. Sci., Serie D, 272:2459.
Caine, N. G., 1976, "Time Separation Pitch and the Dolphin's
    Sonar Discrimination of Distance", Master's thesis
    San Diego State University.
Caldwell, D. K., and Caldwell, M. C., 1966, Observation on the
    distribution, coloration, behavior and audible sound
    production of the spotted dolphin, Stenella plagiodon
    COPE, Los Angeles County Mus., Contrib. in Sci., 104:1.
Caldwell, D. K., and Caldwell, M. C., 1970, Echolocation-type
    signals by two dolphins genus Sotalia, Quart. J. Fla. Acad.
    Sci., 33:124.
Caldwell, D. K., and Caldwell, M. C., 1971, Sounds produced by two
    rare cetaceans stranded in Florida, Cetology, 4:1.
Caldwell, D. K., and Caldwell, M. C., 1971, Underwater pulsed
    sounds produced by captive spotted dolphins, Stenella
    plagiodon, Cetology, 1:1.
Caldwell, D. K., and Caldwell, M.C., 1972, Senses and communica-
    tion, in: "Mammals of the Sea, Biology and Medicine",
    S. H. Ridgway, ed., C. C. Thomas, Springfield, Ill.
Caldwell, M.C., and Caldwell, D. K., 1967, Intraspecific transfer
    of information via pulsed sound in captive Odontocete
    Cetaceans, in: "Animal Sonar Systems: Biology and Bionics",
    R.-G. Busnel, ed., Lab de Physiol. Acoust. Jouy-en-Josas,
    France.
Caldwell, M. C. & Caldwell, D. K., 1968, Vocalization of naive
    captive dophins in small groups, Science, 159:1121.
Caldwell, M. C., and Caldwell, D. K., 1969, Simultaneous but
    different narrow-band sound emissions by a captive Eastern
    Pacific pilot whale, Globicephala scammoni, Mammalia, 33:505.
Caldwell, M. C., and Caldwell, D. K., 1970, Echolocation-type
    signals by dolphins genus Inia, Sea Frontiers, 15:343.
Caldwell, M. C., and Caldwell, D. K., 1970, Etiology of the
    chirp sounds emitted by the atlantic bottlenosed dolphins:
    a controversial issue, Underwater Naturalist, 6:6.
Caldwell, M. C., and Caldwell, D. K., 1970, Further studies on
    audible vocalizations of the Amazon freshwater dolphin,
    Inia geoffrensis, Contrib. in Sci., 187:1.
Caldwell, M. C., and Caldwell, D. K., 1971, Statistical evidence
    for individual signature whistles in Pacific whitesided
    dolphins, Lagenorhynchus obliquidens, Cetology, 3:1.

Caldwell, M. C., and Caldwell, D. K., 1972, Vocal mimicry in
    the whistle mode by an Atlantic bottlenosed dolphin,
    Cetology, 9:1.
Caldwell, M. C., Caldwell, D. K., and Evans, W. E., 1966,
    Preliminary results of studies on the sounds and associated
    behavior of captive Amazon freshwater dolphins, Inia
    geoffrensis, in: "Proc. 3rd Annu. Conf. Biol. Sonar & Diving
    Mammals", Stanford Res. Inst., Menlo Park, Calif.
Caldwell, M. C., Caldwell, D. K., and Evans, W. E., 1966, Sounds
    and behavior of captive amazon freshwater dolphins, Inia
    geoffrensis, Contr. in Science, L. A. County Museum, 108:1.
Caldwell, M. C., Caldwell, D. K., and Hall, N. R., 1969, An
    experimental demonstration of the ability of an Atlantic
    bottlenosed dolphin to discriminate between whistles of
    other individuals of the same species, Los Angeles County
    Mus. Nat. Hist. Found., 6:35.
Caldwell, M. C., Caldwell, D. K., and Hall, N. R., 1970, An
    experimental demonstration of the ability of an Atlantic
    bottlenosed dolphin to discriminate between playbacks of
    recorded whistles of conspecifics, in: "Proc. of the 7th
    Annu. Conf. on Biol. Sonar & Diving Mammals", Menlo Park,
    Calif.
Caldwell, D. K., Caldwell, M. C., and Miller, J. F., 1969, Three
    brief narrow-band sound emissions by a captive male
    Risso's dolphin, Grampus griseus, Bull. S. Calif. Acad.
    Sci., 68:252.
Caldwell, M. C., Caldwell, D. K. and Miller, J. F., 1970,
    Statistical evidence for individual signature whistles in the
    spotted dolphin, Stenella plagiodon, Los Angeles County
    Mus. Nat. Hist. Found., Tech Rept., 7:45
Caldwell, M. C., Caldwell, D. K., and Miller, J. F., 1973, Statis-
    tical evidence for individual signature whistles in the spotted
    dolphin, Stenella plagiodon, Cetology, 16:1.
Caldwell, D. K., Caldwell, M. C., and Rice, D. W., 1966, Behavior
    of the sperm whale, Physeter catodon L., in: "Whales
    Dolphins, and Porpoises", K. S. Norris, ed., Univ. of
    California Press, Berkeley.
Caldwell, M. C., Caldwell, D. K., and Turner, R. H., 1970, Statis-
    tical analysis of the signature whistle of an Atlantic bot-
    tlenose dolphin with correlations between vocal changes and
    level arousal, Los Angeles Co., Calif. Mus. Nat. Hist., 8:1.
Caldwell, M. C., Hall, N. R., and Caldwell, D. K., 1971, Ability
    of an atlantic bottlenosed dolphin to discriminate between,
    and potentially identify to individual, the whistles
    of another species, the spotted dolphin, Cetology, 6:1.
Caldwell, M. C., Hall, N. R., and Caldwell, D. K., 1971, Ability
    of an Atlantic bottlenosed dolphin to discriminate between,
    and respond differentially to whistles of eight conspecifics,
    in: "Proc. 8th Annu. Conf. Biol. Sonar & Diving Mammals",
    Menlo Park, Calif.

Caldwell, D. K., Prescott, J. H., and Caldwell, M. C., 1966, Production of pulsed sound by the pigmy sperm whale, Kogia breviceps, Bull. S. Calif. Acad. Sci., 65:245.

Caldwell, M. C., Hall, N. R., Caldwell, D. K., and Hall, H. I., 1971, A preliminary investigation of the ability of an Atlantic bottlenosed dolphin to localize underwater sound sources, Marineland Res. Lab., Techn. Rept. 6.

Chapman, S., 1968, Dolphins and multifrequency, multiangular images, Science, 160:208.

Cherbit, G., and Alcuri, G., 1978, Etude de la propogation des vibrations a travers le rostre de Sotelia teuzii (cetacea) par inferometric holographique, C. R. Acad. Sci. Paris, 286:607.

Corcella, A. T., and Green, M., 1968, Investigation of impulsive deep-sea noise resembling sounds produced by a whale, Jour. Acoust. Soc. Amer., 44:483.

Cummings, W. C., and Philippi, L. A., 1970, Whale phonations in repetitive stanzas, NUC Techn. Publ. 196, San Diego, Calif.

Cummings, W. C., and Thompson, P. O., 1968, Self-noise of a deep submersible in a bioacoustic investigation off Catalina Island, California, Jour. Acoust. Soc. Amer., 44:1742.

Cummings, W. C. and Thompson, P. O., 1971, Gray whales, Eschrichtus robustus, avoid the underwater sounds of killer whales, Orcinus orca, Fishery Bull., 69:525.

Cummings, W. C., and Thompson, P. O., 1971, Underwater sounds from the blue whale, Balaenoptera musculus, Jour. Acoust. Soc. Amer., 50:1193.

Cummings, W. C. and Thompson, P. O., 1971, Bioacoustics of marine mammals, R. V. Hero Cruise 70-3, Antarctic J. U. S., 6:158.

Cummings, W. C., Fish, J. F., and Thompson, P. O., 1972, Sound production and other behavior of southern right whale, Eubalena glacialis, Transact. of the San Diego Soc. Natur. Hist., 17:1.

Cummings, W. C., Thompson, P. O., and Cook, R., 1967, Sound production of migrating gray whales, Eschrichtius gibbosus ERXLEBEN, 74th Meeting of the Acoustical Soc. of America.

Cummings, W. C., Thompson, P. O., and Cook, R., 1968, Underwater sounds of migrating gray whales, Eschrichtius glaucus COPE, Jour. Acoust. Soc. Amer., 44:1278.

Cummings, W. C., Fish, J. F., Thompson, P. O., and Jehl, J. R., 1971, Bioacoustics of marine mammals off Argentina, R. V. Hero Cruise 71-3, Antarct. J. U. S., 6:266.

Dailly, M., 1971, The primary cochlear nuclei in the Amazon dolphin Inia geoffrensis, in: "Investigations on Cetacea: Vol. III", G. Pilleri, ed., Bentelli, Berne.

Dailly, M., 1972, Contribution to the study of the cochlear apparatus in dolphins, in: "Investigations on Cetacea, Vol. II", G. Pilleri, ed., Bentelli, Berne.

Defran, R. H., and Caine, N. G., 1976, Periodicity pitch difference
    limens in the bottlenose dolphin (Tursiops truncatus), Jour.
    Acoust. Soc. Amer., 60: (Supplement 1) S5 (Abstract).

Diercks, K. J., 1972, Biological sonar systems: a bionics survey,
    Applied Research Laboratories, ARL-TR-72-34, Austin, Texas.

Diercks, K. J., 1974, A collection of translations of foreign
    language papers on the subject of biological sonar systems,
    Applied Res. Lab. ARL-TR-74-9, Austin, Texas.

Diercks, K. J., and Evans, W. E., 1975, Delphinid sonar: pulse
    wave and simulation studies, NUC Techn. Publ. 175, San Diego.

Diercks, K. J., and Hickling, R., 1967, Echoes from hollow aluminum
    spheres in water, Jour. Acoust. Soc. Amer., 41:380.

Diercks, K. J., and Trochta, R. T., 1972, Animal sonar: measurements
    and meaning, Jour. Acoust. Soc. Amer, 51:133.

Diercks, K. J., Trochta, R. T., and Evans, W. E., 1973, Delphinid
    sonar: measurement and analysis, Jour. Acoust. Soc. Amer.,
    54:200.

Diercks, K. J., Trochta, R. T., Greenlaw, R. L. and Evans, W. E.,
    1971, Recording and analysis of dolphin echolocation signals,
    Jour. Acoust. Soc. Amer., 49:1729.

Dmitrieva, I. L., Krushinskaya, N. L., Bel'kovich, V. M., and
    Shurkhal, A. V., 1972, Echolocational discrimination of
    the shape of different subjects by dolphins, in: "Proc. of the
    Abstracts of the 5th All-Union Conference on Marine Mammals",
    Makhachkala, Part 2:76.

Dormer, K. J., 1974, The mechanism of sound production and measure-
    ment of sound processing in delphinid cetaceans, Ph.d.
    dissertation, University of California, Los Angeles.

Dormer, K. J., 1979, Mechanisms of sound production and air
    recylcing in delphinids: Cineradiographic evidence, Jour.
    Acoust. Soc. Amer., 65:229.

Dral, A. D. G., 1972, Aquatic and aerial vision in the bottlenosed
    dolphin Neth. J. Sea Res, 5:510.

Dreher, J. J., 1966, Cetacean communication: small-group
    experiment, in: "Whales, Dolphins, and Porpoises", K. S.
    Norris, ed., Univ. Calif. Press, Berkeley.

Dreher, J. J., 1967, Bistatic target signatures and their acoustic
    recognition: a suggested animal model, in: "Marine Bio
    Acoustics", W. N. Tavolga, ed., Pergamon Press, N. Y.

Dreher, J. J., 1969, Acoustic holographic model of cetacean
    echolocation, in: "Acoustical Holography, Vol. I", A. E.
    Metherell, ed., Plenum Press, New York.

Dubrovskiy, N. A., 1972, Discrimination of objects by dolphins
    using echolocation, "Report of the 5th All-Union Conference
    on Studies of Marine Mammals, Part 2, Makhachkala", as cited
    in: Ayrapet'yants, E. Sh. and Konstantinov, A. I., 1974,
    "Echolocation in Nature", Nauka, Leningrad (English translation
    JPRS 63328-2).

Dubrovskiy, N. A., and Krasnov, P. S., 1971, Discrimination of elastic spheres according to material and size by the bottlenose dolphin, Trudy Akusticheskogo Institute, 17, as cited in: Bel'kovich, V. M., and Dubrovskiy, N. A., 1976, "Sensory Bases of Cetacean Orientation", Nauka, Leningrad (English translation JPRS L/7157).

Dubrovskiy, N. A., and Titov, A. A., 1975, Echolocation discrimination by bottlenose dolphins (Tursiops truncatus) of spherical targets which differ simultaneously in dimensions and materials, Akusticheskiy Zhurnal, 21(3):469.

Dubrovskiy, N. A., and Zaslavsky, G. L., Dolphin echolocation Akusticheskiy Zhurnal, 3 (English Translation JPRS 65777).

Dubrovskiy, N. A., and Zaslavsky, G. L., 1973, Temporal structure and directivity of sound emission by the bottlenosed dolphin, in: "Proceedings of the 8th All-Union Acoustical Conference".

Dubrovskiy, N. A., and Zaslavski, G. L., 1975, Role of the skull bones in the space-time development of the dolphin echolocation signal, Sov. Phys. Acoust., 21:409 (Translation: American Institute of Physics).

Dubrovskiy, N. A., Krasnov, P. S., and Titov, A. A., 1970, On the emission of echo-location signals by the Azov Sea harbor porpoise, Soviet Physics-Acoustics, 16:444.

Dubrovskiy, N. A., Krasnov, P. S. and Titov, A. A., 1970, On the question of the emission of ultrasonic ranging signals by the common porpoise, Akusticheskiy Zhurnal, 16:521 (English translation JPRS 52291).

Dubrovskiy, N. A., Krasnov, P. S., and Titov, A. A., 1971, Discrimination of solid elastic spheres by an echolocating porpoise, Tursiops truncatus, in: "Proc. 7th International Acoust.", Budapest, 25(3):533.

Dubrovskiy, N. A., Titov, A. A., Krasnov, P. S., Babkin, V. P. Lekomtsev, V. M., and Nikolenko, G. V., 1970, Investigation of the emission capacity of the Black Sea Tursiops truncatus echolocation apparatus, Trudy Akusticheskogo Instituta, 10:163.

Dudok van Heel, W. H., 1966, Navigation in Cetacea, in: "Whales Dolphins and Porpoises", K. S. Norris, ed., U. of Calif. Press, Berkeley.

Dunn, J. L., 1969, Airborn measurements of the acoustic characteristics of a sperm whale, Jour. Acoust. Soc. Amer., 46:1052.

Dziedzic, A., 1968, Quelques performances des systemes de detection par echos des chauves souris et des delphinidae, Rev. Acoust., 1:23.

Dziedzic, A., 1971, Performances dans la detection de cibles par echolocation par le Marsouin Phocoena phocoena et etude analytique des signaux mis en oeuvre dans ce processus, These, Ecole Pratique des Hautes Etudes.

Dziedzic, A., 1971, Theorie et applications de l'acoustique sousmarine Ecole Super. des Techn. avancees, 16:1.

Dziedzic, A., 1972, Les sonars biologiques, La Recherche, 3:315.

Dziedzic, A., and Alcuri, G., 1977, Acoustic recognition of forms and characteristics of the sonar signals of Tursiops truncatus, C. R. Acad. Sci. Paris, 285:981.

Dziedzic, A., and Levy, J. C., 1973, Description et mise en oeuvre d'une base de triangulation destinee a l'etude en mer des signaux de Dauphins, 4°Coll. Traitement du signal et ses applications, Nice, 4:585.

Dziedzic, A., Escudie, B., and Hellion, A., 1975, Realistic methods for analysis of biological sonar signals, Ann. Telecomm., 30:270.

Dziedzic, A., Jeanny, M., and Levy, J. C., 1975, Resultats d'une etude sur les caracteristiques spatio-temporelles et spectrales des signaux acoustiques emis par un dauphin l'approche sonar d'une cible, in: "Colloque Nat. Sur le Traitement du Signal et ses Applications", Nice.

Dziedzic, A., Chiollaz, M., Escudie, B., and Hellion, A., 1977, Some properties of low frequency sonar signals of the dolphin Phoecena phoecena, Acustica, 37:258.

Dziedzic, A., Escudie, B., Guillard, P., and Hellion, A., 1974, Evidence of tolerance to the effect of Doppler in the sonar emissions of Delphinus delphis, C. R. Acad. Sci. Paris, 279:1313.

Dziedzic, A., Escudie, B., Hellion, A., and Vial, C., 1969, Resultats preliminaires d'une etude de certains sonars biologiques - Analyse spectrale de signauz de 7 especes de Cetaces odontocetes, 2°Coll. traitement du signal et ses applications, Nice, 2:785.

Eberhardt, R. L. and Evans, W. E., 1962, Sound activity of the California gray whale, Eschrichtius glaucus, Jour. Aud. Eng. Soc., 10:324.

Erulkar, S. D., 1972, Comparative aspects of spatial localization of sound, Physiological Reviews, 52:237.

Escudie, B., 1972, Etat actuel des travaux sur les "sonars biologiques", Comparaison des proprietes de ces systemes avec les artificiels, Perspectives, I. C. P. de Lyon, 2:27.

Escudie, B., Hellion, A., Dziedzic, A., 1971, Results from studies of air and marine sonars through signal processing and spectral analysis, in: "Proc. Troisieme Colloque sur le Traitement Signal et ses Applications", 3:533.

Evans, W. E., 1967, Discussion, in: "Animal Sonar Systems, Biology and Bionics", R.-G. Busnel, ed., Laboratoire de Physiology Acoustique, Jouy-en-Josas, France.

Evans, W. E., 1967, Vocalizations among marine mammals, in: "Marine Bio-Acoustic II", W. N. Tavolga, ed., Pergamon Press, New York.

Evans, W. E., 1969, Marine mammal communication: Social and ecological factors, in: "The Biology of Marine Mammals", H. T. Andersen, ed., Academic Press, New York.

Evans, W. E., 1973, Echolocation by marine delphinids and one species of fresh-water dolphin, Jour. Acoust. Soc. Amer., 54:191.

Evans, W. E. and Bastian, J., 1969, Marine mammal communication: social and ecological factors, in: "The Biology of Marine Mammals", H. T. Anderson, ed., Academic Press, N. Y.

Evans, W. E., and Herald, E. S., 1970, Underwater calls of a captive Amazon manatee, Trichechus inunguis, J. Mammal, 51:820.

Evans, W. E., and Maderson, P. F. A., 1973, Mechanisms of sound production in delphinid cetaceans: A review and some anatomical considerations, Amer. Zool., 13:1205.

Evans, W. E., and Powell, B. A., 1967, Discrimination of different metallic plates by an echolocating delphinid, in: "Animal Sonar Systems, Biology and Bionics", R.-G. Busnel, ed., Laboratoire de Physiologie Acoustique, Jouy-en-Josas, France.

Fadeyeva, L. M., 1973, Discrimination of spheres with various echosignal structure by dolphins, in: "Proceedings of the 8th All-Union Acoustical Conference", 1:134.

Fay, R. R., 1974, Auditory frequency discrimination in vertebrates, Jour. Acoust. Soc. Amer., 56:206.

Filimonoff, I. N., 1966, On the so-called rhinencephalon in the dolphin, J. Hirnforsch, 8:1.

Fish, J. F., and Lingle, G. E., 1977, Responses of spotted porpoises (Stenella attenuata) to playback of distress (?) sounds of one of their kind, in: "Proceedings (Abstracts) Second Conference on the Biology of Marine Mammals", San Diego.

Fish, J. F., and Turl, C. W., 1976, "Acoustic source levels of four species of small whales", NUC Techn. Publ. 547, San Diego.

Fish, J. F., and Vania, J. S., 1971, Killer whale, Orcinus orca, sounds repel white whales, Delphinapterus leucas, Fishery Bull., 69:531.

Fish, J. F., and Winn, H. E., Sounds of marine animals, in: "Encyclopedia of Marine Resources", F. E. Firth, ed., Van Nostrand Reinhold Co.

Fish, J. F., Johnson, C. S., and Ljungblad, D. K., 1976, Sonar target discrimination by instrumented human divers, Jour. Acoust Soc. Amer., 59:602.

Fish, J. F., Sumich, J. L. and Lingle, G. E., 1974, Sounds produced by the gray whale, Eschrichtius robustus, Mar. Fish Rev., 36:38.

Fitzgerald, J. W., 1973, "A qualitative model of the dolphin sonar, Part II: The high-frequency echo-location apparatus", Technical Report 117, Office of Naval Res., Acoustics Progr. Off., Arlington, Virginia.

Flanigan, N. J., 1972, The central nervous system, in: "Mammals of the Sea: Biology and Medicine", S. H. Ridgway, ed., C. C. Thomas, Springfield, Ill.

Fleischer, G., 1973, On structure and function of the middle ear in the bottlenosed dolphin (Tursiops truncatus), in: "Proc. 9th Ann. Conf. Biol. Sonar and Diving Mammals", Stanford Res. Inst. Press, Menlo Park, Calif.

Fleischer, G., 1973, Structural analysis of the tympanicum complex in the bottle-nosed dolphin (Tursiops truncatus), Jour. Aud. Res., 13:178.

Fleischer, G., 1975, Uber das spezialisierte Gehororgan von Kogia Breviceps (odontoceti), Z.f. Saugetierkunde Bd. 40, H.2, S:89.

Fleischer, G., 1976, Hearing in extinct cetaceans as determined by cochlear structure, Journal of Paleontology, 50:133.

Fleischer, G., 1976, On bony microstructures in the dolphin cochlea, related to hearing, N. Jb. Geol. Palaont., 151:166.

Fleischer, G., 1976, Uber Beziehungen Zwischen Hovermogen und Schadelbau bie Walen, Saugetierkundliche Mitteilungen: 48.

Fleischer, G., 1976, Uber Die Verankerung des Stapes im Ohr der Cetacea und Sirenia, Z.f. Saugetierkunde Bd. 41, H.5, S:304.

Fleischer, G., 1978, Evolutionary principles of the mammalian middle ear, in: "Advances in Anatomy, Embryology and Cell Biology; Springer Verlag, Berlin.

Ford, J. K. B., and Fischer, H. D., 1978, Underwater acoustic signals of the narwhal (Monodon monoceros), Can. J. Zool., 56:552.

Fraser, F. C., 1973, Record of a dolphin (Sousa teuszii) from the coast of Mauritania, Transact. N. Y. Acad. Sci., 35:132.

Gales, R. S., 1966, Pickup analysis and interpretation of underwater acoustic data, in: "Whales, Dolphins and Porpoises", K. S. Norris, ed., U. of Calif. Press, Berkeley.

Gallien, C. L., Chalumeau-le-Foulgoc, M. T., and Fine, J. M., 1967, Les proteines serigues de Delphinus delphis Linne (Cetace Odontocete) C. R. Acad. Sci. Paris, 264:1359.

Gallien, C. L., Chalumeau-le-Foulgoc, M. T., and Fine, J. M., 1970, Comparative study of serum proteins in four dolphin species, Comp. Biochem. Physiol., 37:375.

Gapich, L. I. and Supin, Y. A., 1974, The activity of single neurons of the acoustical region of the cerebral cortex of the porpoise Phocoena phocoena, J. Evol. Biochem. Physiol., 10:182.

Gentry, R. L., 1967, Underwater auditory localization in the California sea lion (Zalophus californianus), Jour. Aud. Res., 7:187.

Giraud-Sauveur, D., 1969, Biophysical research on the ossicles of cetaceans, Extrait de Mammalia, 33:285.

Giraud-Sauveur, D., and Miloche, M., 1968, Sur la structure particuliere des os de l'oreille moyenne des cetaces odontocetes, J. Microscop., 7:1098.

Giro, L. R. and Dubrovskiy, N. A., 1972, Correlation between
     recurrence frequency of delphinid echolocation signals and
     difficulty of echoranging problem, in: "Marine Instrument
     Building Series, 2:84, as cited in: Vel'min V. A., 1975,
     Target detection by the bottlenose dolphin under artificial
     reverberation conditions, in: "Marine Mammals, Proceedings
     of the Sixth All-Union Conference on the Study of Marine
     Mammals", G. B. Agarkov, ed., Nauka Dumka, Kiev (English
     translation JPRS L/6049).

Giro, L. R. and Dubrovskiy, N. A., 1973, On the origin of low-
     frequency component of the echoranging pulsed signal of
     dolphins in: "VIII-aya Vsesoyuznaya Akusticheskaya
     Konferentsiya", (Eighth All-Union Acoustical Conference)
     Moscow.

Giro, L. R., and Dubrovskiy, N. A., 1974, The possible role
     of supracranial air sacs in the formation of echo-ranging
     signals in dolphins, Akusticheskiy Zhurnal, 20(5):706.

Golubkov, A. G., and Ivanenko, Yu. V., 1970, Study of the range
     resolution ability of the dolphin's echolocator, in: "Proc. of
     the 23rd Scientific-Technical Conference of LIAP",

Golubkov, A. G., Ershova, I. V., Korolev, I. V. and Malyshev,
     Yu. A., 1972, On energetical parameters of the Black
     Sea bottlenosed dolphin echolocational apparatus, Trudy
     LIAP 76:9.

Golubkov, A. G., Korolev, V. I., Antonov, V. A., and Ignat'eva,
     E. A. 1975, Comparison of dolphin echolocation signals with
     results of optimal signal calculations, Doklady Akad. Nauk SSSR
     223:1251.

Golubkov, A. G., Zworykhin, V. N., Ershova, I. V., Korolev, V. I.,
     Burdin, V. I., and Malyshev, Ju. A., 1969, Some experimental
     data and prospects for investigation of the analyzing capa-
     city in dolphins, Trudij Leningr. Inst. Aviats., 64:128.

Gracheva, M. S., 1971, A contribution to the structure of Larynx
     in Tursiops truncatus, Rev. de Zool., Moscow, 50:1539.

Grinnel, A. D., 1967, Mechanisms of overcoming interference in
     echolocating animals, in: "Animal Sonar Systems, Biology and
     Bionics", R.-G. Busnel, ed., Laboratoire de Physiologie
     Acoustique, Jouy-en-Josas, France.

Gruenberger, H. B., 1970, On the cerebral anatomy of the Amazon
     dolphin, Inia geoffrensis, in: "Investigations on Cetacea
     V. II", G. Pilleri, Bentelli AG., Berne.

Gurevich, V. S., 1968, Experience of study of dolphin's capabili-
     ties (Delphinus delphis) for echolocation discrimination of
     geometrical figures, (Theses of report), "Scientific
     Conference of Moscow State Lomonosov University, Moscow.

Gurevich, V. S., 1969, Echolocation discrimination of geometric
     figures in the dolphin, Delphinus delphis, Moscow, Vestnik
     Moskovskoga Universiteta, Biologiya, Pochovedeniye, 3:109
     (English Translation JPRS 49281).

Gurevich, V. S., 1972, Morpho-functional studying of the upper
    skull respiratory tracts of the Common dolphin (Delphinus
    delphis), in: "Conference on using Mathematical Methods
    and Computing in Mathematics and Bionics", Leningrad.

Gurevich, V. S., 1973, Experience of the plastical reconstruction
    of the upper skull respiratory tract of the Common dolphin
    (Delphinus delphis), Journal Ontogenez: 78.

Gurevich, V. S., 1973, Histological study of the sound
    production organs in the Lagenorhynchus obliquidens, in:
    "Proceedings of the Eighth All-Union Acoustical Conference",
    Leningrad.

Gurevich, V. S., 1973, Possibilities of interaction of the larynx
    and upper skull respiratory tracts in the process of sound
    production in the dolphins, in: "Material of the Fourth
    All-Union Conference on Marine Mammals", Makhachkala: 23-25.

Gurevich, V. S., and Evans, W. E., 1976, Echolocation discrimina-
    tion of complex planar targets by the Beluga whale
    (Delphinapterus leucas), Jour. Acoust. Soc. Amer., 60(1):5.

Gurevich, V. S., and Evdokimov, V. N., 1973, Rentgenological
    investigation of the lungs and respiratory tracts of the
    Common dolphin (Delphinus delphis), in: "Materials of the
    Fourth All-Union Conference Marine Mammals", Makhachkala, 6:14.

Gurevich, V. S., and Korol'kov, Yu. I., 1973, A rentgenological
    study of respirative act in Delphinus delphis, Zool. Jour.,
    52(5):786.

Hall, J. G., 1967, Hearing and primary auditory centers of the
    whales, Acta Otolaryngol., Suppl., 224:244.

Hall, J. D., and Johnson, C. S. 1972, Auditory thresholds of a
    killer whale Orcinus orca (Linnaeus), Jour. Acoust.
    Soc. Amer., 51:515.

Hammer, C. E., 1977, Biosonar report: An experimental analysis
    of salient target characteristics for the echo-locating
    porpoise (Tursiops truncatus) (Technical Report 7709-1),
    SEACO, Incorporated, Kailua, Hawaii.

Hammer, C. E., 1978, Echo-recognition in the porpoise (Tursiops
    truncatus): An experimental analysis of salient target
    characteristics, Naval Ocean Systems Center, San Diego,
    Tech. Rep. 192.

Hammer, C., and Au, W. W. L., 1978, Target recognition via
    echolocation by an Atlantic bottlenose porpoise (Tursiops
    truncatus), Jour. Acoust. Soc. Amer., 64(1):587.

Hansen, I. A., and Cheah, C. C., 1969, Related dietary and
    tissue lipids of the sperm whale, Comp. Biochem. Physiol.,
    31:757.

Hellion, A., Guillard, P., Escudie, B., and Dziedzic, A.,
    1974, Etude des signaux impulsifs utilises par les sonars
    biologiques, Coll. s. les methodes d'etudes et simulation
    des chocs, Lyon.

Herman, L. M., 1975, Interference and auditory short-term
    memory in the bottlenosed dolphin, Animal Learning
    & Behavior., 3:43.
Herman, L. M., and Arbeit, W. R., 1971, Auditory frequency
    discrimination from 1-36 kHz in Tursiops truncatus,
    in: "Proc. 8th Annu. Conf. on Biol. Sonar and Diving
    Mammals", Stanford Res. Inst., Menlo Park, Calif.
Herman, L. M. and Arbeit, W. R., 1972, Frequency difference
    limens in the bottlenose dolphin: 1-70 kHz, Jour. Aud. Res.,
    12:109.
Herman, L. M., and Arbeit, W. R., 1973, Stimulus control
    and auditory discrimination learning sets in the
    bottlenose dolphin, J. exp, Analys. Behav., 19:379.
Herman, L. M., and Gordon, J. A., 1974, Auditory delayed
    matching in the bottlenose dolphin, J. Exp. Analys.
    Behav., 21:19.
Herman, L. M., Beach, F. A., Pepper, R. L., and Stalling, R. B.,
    1969, Learning-set formation in the bottlenose dolphin,
    Psychon. Sci., 14:98.
Herman, L. M., Peacock, M. F., Yunker, M. P., and Madsen,
    C. J., 1975, Bottlenosed dolphin: Double split
    pupil yields equivalent aerial and underwater duirnal
    acuity, Science, 189:650.
Herald, E. S., Brownell, R. L., Frye, R. L., Morris, E. J.,
    Evans, W. E., and Scott, A. B., 1969, Blind river
    dolphin: First side-swimming cetacean, Science,
    166:1408.
Hollien, H., Hollien, P., Caldwell, D. K., and Caldwell, M. C.,
    1976, Sound production by the Atlantic bottlenosed
    dolphin, Tursiops truncatus, Cetology, 26:1.
Ivanova, M. P., and Kurganskiy, N. A., 1973, A dolphin's
    trajectory of motion in the process of discriminating
    spherical and cylindrical objects, IV-oe Vsesoyuznoye
    soveshchaniye po Bionike, Moscow, 2:106.
Jacobs, D., 1972, Auditory frequency discrimination in the
    Atlantic bottlenose dolphin, Tursiops truncatus, Montague:
    A preliminary report, Jour. Acoust. Soc. Amer., 52:696.
Jacobs, D. W., and Hall, J. D., 1972, Auditory thresholds of a
    freshwater dolphin, Inia geoffrensis Blainville, Jour. Acoust.
    Soc. Amer., 51:530.
Jacobs, M., 1971, "Talking" through the nose. Sound and communica-
    tion in whales and dolphins, Anim. Kingdom, 74:2.
Jansen, J., and Jansen, J. K. S,. 1969, The nervous system of
    cetecea, in: "The Biology of Marine Mammals", H. T. Anderson,
    ed., Acad. Press, New York.
Jerison, H. J., 1973, "Evolution of the Brain and Intelligence",
    Acad. Press, New York.
Johnson, C. S., 1966, Auditory thresholds of the bottlenosed
    porpoise, Tursiops truncatus (Montague), N. O. T. S. TP
    4178.

Johnson, C. S., 1967, Discussion to paper by Evans and Powell,
    in: "Animal Sonar Systems, Biology and Bionics", R.-G.
    Busnel, ed., Laboratoire de Physiologie Acoustique, Jouy-en-
    Josas, France.

Johnson, C. S., 1967, Sound detection thresholds in marine mammals,
    in: "Marine Bio-Acoustics, Proc. of the Second Symposium on
    Marine Bio-Acoustics, New York", W. N. Tavolga, ed., Pergamon
    Press, New York.

Johnson, C. S., 1968, Masked pure tone thresholds in the bottle-
    nosed porpoise, Jour. Acoust. Soc. Amer., 44:965.

Johnson, C. S., 1971, Auditory masking of one pure tone by
    another in the bottlenosed porpoise, Jour. Acoust. Soc.
    Amer., 49:1317.

Johnson, R. A., and Titlebaum, E. L., 1976, Energy spectrum
    analysis: A model of echolocation processing, Jour. Acoust.
    Soc. Amer. 60:484.

Kadnadzey, V. V., Kreychi, S. A., Kakhalkina, E. N., Nikolenko,
    G. V., and Titov, A. A., 1975, Morskiye Mlekopitayushchiye,
    in: "Sixth All-Union Conference on the Study of Marine
    Mammals", G. B. Agarkov, ed., (English translation JPRS
    L6049-1).

Kanazu, R., Morii, H., and Fukuhara, T., 1969, Studies on the
    little toothed whales in the west sea area of Kyushu - XVII,
    about higher branched chain fatty acids in head oil of the
    little toothed whales - I., Bull. Fac. Fish. Nagasaki Univ.
    28:161.

Karol, R., Litchfield, C., Caldwell, D. K., and Caldwell, M. C.,
    1978, Compositional topography of the melon and spermaceti
    organ lipids in the pygmy sperm whale Kogia breviceps:
    Implications for echolocation, Mar. Biol., 47:115.

Kasuya, T., 1973, Systematic consideration of recent toothed whales
    based on the morphology of tympano-periotic bone, Sci, Rept.
    Whales Res. Inst., 25:103.

Kaufman, B. W., Siniff, D. B., and Reichle, R. A., 1972, Colony
    behavior of Weddell seals, Leptonychotes weddelli, at Hutton
    Cliffs, Antartica, in: "Symposium on the Biology of the
    Seal", G. B. Farquhar, ed., Univ. of Guelph, Ontario.

Kenshalo, D. R., 1967, Discussion, in: "Animal Sonar Systems,
    Biology and Bionics", R.-G. Busnel, ed., Laboratoire de
    Physiologie Acoustique, Jouy-en-Josas, France.

Kesarev, V. S., Trykova, O. V., and Malofeyeva, L. I., 1975,
    Structural prerequisites for corticalization of the dolphin's
    acoustic analyzer, in: "Morskiye Mlekopitayushchiye", G. B.
    Agarkov, ed., Izdatel'stvo Naukova Dumka, Kiev.

Khakhalina, E. N., 1975, Emotional signals in the dolphin's
    communication system, in: "Morskiye Mlekopitayushchiye",
    G. B. Agarkov, ed., Isdatel'stvo Naukova Dumka, Kiev.

Khomenko, B. G., 1969, Distinctions in structure of the peripheral nervous system of the dolphin's melon (Tursiops truncatus Ponticus Barabash), in: "Tezisy II vsesoyuznogo simpoziuma mododykh uchenykh", Sevastopol.

Khomenko, B. G., 1970, Some distinctions of histological structure and innvervation of the rostrum of the Black Sea dolphins, Bionika, 4:70.

Khomenko, B. G., 1970, The histostructure and innvervation of the sound apparatus -nasal sacs - in dolphins, Akad. Nauk. URSS, Dopovid, 32:83.

Khomenko, B. G., 1973, A contribution to the morphology of I-VIII pairs of cranial nerves in dolphins, Zoologicheskiy Zhurnal, 52(3):407.

Khomenko, B. G., 1973, Morphological basis for the echolocation properties of dolphins, Bionika, 7:60.

Khomenko, B. G., 1974, A morphological and functional analysis of the structure of the supercranial respiratory passage as a possible echolocation signal generator in the dolphin, Bionika, 8 (English translation JPRS 63492).

Khomenko, B. G., 1975, Comparative anatomical analysis of the structure and innervation of nasolabial region in certain mammals and man, Vestnik Zoologii, 5:39.

Khomenko, B. G., 1975, Morphological characteristics of the respiratory and suckling reflexes in dolphin embryogenesis, in: "Morskiye Mlekopitayushchiye", G. B. Agarkov, ed., Izdatel'stvo Naukova Dumka, Kiev.

Khomenko, B. G., and Khadzhinskiy, V. G., 1974, Morphological and functional principles underlying cutaneous reception in dolphins, Bionika, 8:106.

Kibblewhite, A. C., Denham, R. N., and Barnes, D. J., 1967, Unusual low-frequency signals observed in New Zealand waters, Jour. Acoust. Soc., Amer., 41:644.

Kinne, 0., 1975, Orientation in space: Animals, marine, in: "Marine Ecology Vol. 2 - Physiological Mechanisms", O. Kinne, ed., Wiley, London.

Kiselev, V. I., and Mrochkov, K. A., 1975, Products of dissolution of cetacean connective tissue collagen and use thereof, in: "Morskiye Mlekopitayushchiye", G. B. Agarkov, ed., Izdatel'stvo Naukova Dumka, Kiev.

Konstantinov, A. I., Mel'nikov, N. F., and Titov, A. V., 1968, On dolphin ability of identifying objects, in: "Tezisy Dokladov na II-oy Ukrainskoy Respubliknaskoy Konferentsii po Bionike", Kiev.

Korolev, L. D., Lipatov, N. V., and Resvov, R. N., Savel'ev, M.A., and Flenov, A. B., 1973, Investigation of the spatial directivity diagram of the dolphin sound emitter, in: "Proceedings of the 8th All-Union Acoustical Conference", Leningrad.

Koshovoy, V. V., and Mykhaylovs'kyy, V. M., 1972, The operating
    principle of the echolocating system of marine mammals,
    Dopovidi Akad. Nauk Ukrayins'kayi RSR Ser. A Fizyko-tekhn.
    ta matemat. Nauky, Ukranian 12:1097 (English translation
    JPRS 58344).

Kozak, V. A., 1974, Receptor zone of the video-acoustic system of
    the sperm whale (physeter Catodon L., 1758), Fiziologichnyy
    Zhurnal Academiy Nauk Ukraynskoy RSR 120 (English transla-
    tion: JPRS 65017).

Kruger, Lawrence, 1966, Specialized features of the cetacean
    brain, in: "Whales, Dolphins, and Porpoises", K. S. Norris,
    ed., Univ. of California Press, Berkeley.

Krushinskii, L. V., Dashevskii, B. A., Krushinskaya, N. L., and
    Dmitrieva, I. L., 1972, A study of the ability of the
    dolphin Tursiops truncatus (Montague) to operate with the
    empirical dimensionality of geometric figures, Doklady
    Akademii Nauk SSR, 204:755 (Translation: Plenum Publishing
    Company, New York, 1972).

Ladygina, T. F., and Supin, A. Ya., 1970, The acoustic projection
    in the dolphin cerebral cortex, Fiziol. Zh. SSSR im. I. M.
    Sechenova, 56:1554.

Ladygina, T. F., and Supin, A. Ya., 1977, Localization of sensory
    projection zones in the cerebral cortex of the bottlenosed
    dolphin, Tursiops truncatus, Zurnal Evolyutsionnoy
    Biokhimii i Fiziologii, 6:712.

Leatherwood, J. S., Johnson, R. A., Ljungblad, D. K., and Evans,
    W. E., 1977, "Broadband measurements of underwater acoustic
    target strengths of panels of tuna nets", NOSC TR 126, San
    Diego.

Lekomtsev, V. M., and Titov, A. A., 1974, Procedures for studying
    the echolocation apparatus of dolphins, Bionika, 8:83
    (English Translation JPRS 63492).

Lende, R. A., and Walker, W. I., 1972, An unusual sensory area
    in the cerebral neocortex of the bottlenose dolphin, Tursiops
    truncatus, Brain Res., 45:555.

Levenson, C., 1974, Source level and bistatic target strength
    of the sperm whale (Physeter catodon) measured from an
    oceanographic aircraft, Jour. Acoust. Soc. Amer., 55:1100.

Lilly, J. C., 1966, Sonic-ultrasonic emissions of the bottlenose
    dolphin, in: "Whales, Dolphins, and Porpoises", K. S. Norris,
    ed., Univ. Calif. Press, Berkeley.

Lilly, J. C., 1967, Dolphin vocalization, in: "Brain Mechanisms
    Underlying Speech and Language", Grune and Stratton,
    New York.

Lilly, J. C., 1967, Dolphin's vocal mimicry as a unique ability
    and a step toward understanding, in: "Research in Verbal
    Behavior and Some Neurophysiological Implications",
    Academic Press, New York.

Lilly, J. C., 1967, Mind of the dolphin: a nonhuman intelligence,
    Doubleday and Co., New York.

Lilly, J. C., 1968, Sound production in Tursiops truncatus (bottlenose dolphin), Ann. N. Y. Acad. Sci, 155:320.

Lilly, J. C., Miller, A. M., and Truby, U. M., 1968, Reprogramming of the sonic output of the dolphin sonic burst count matching, Jour. Acous. Soc. Amer., 43:1412.

Lipatov, N. V., and Solntseva, G. N., 1974, Morphological and functional features of the external auditory canal of the common and bottlenose dolphin, Bionika, 8 (English translation JPRS 63492).

Litchfield, C., and Greenberg, A. J., 1974, Comparative lipid patterns in the melon fats of dolphins, porpoises, and toothed whales, Comp. Biochem. Physiol., 478:401.

Litchfield, C., Karol, R., and Greenberg, A. J., 1973, Compositional topography of melon lipids in the Atlantic bottlenose dolphin Tursiops truncatus: implications for echolocation Mar. Biol., 23:165.

Litchfield, C., Ackman, R. G., Sipos, J. C. and Eaton, C. A., 1971, Isovaleroyl triglycerides from the blubber and melon oils of the beluga whale (Delphinapterus leucas), Lipids, 6:674.

Litchfield, C., Kinneman, J., Ackman, R. G., and Eaton, C. A., 1971, Comparative lipid patterns in two freshwater dolphins Inia geoffrensis and Sotalia fluviatilis, Jour. Amer. Oil Chem. Soc., 48:91.

Litchfield, C., Greenberg, A. J., Caldwell, D. K., Caldwell, M. C., Sipos, J. C., and Ackman, R. G., 1975, Comparative lipid patterns in acoustical and nonacoustical fatty tissues of dolphins, porpoises, and toothed whales, Comp. Biochem. Physiol., 508:591.

Livshits, M. S., 1974, Some properties of dolphin hydrolocator from the viewpoint of correlation hypothesis, Biofizika, 19:916 (English translation JPRS 64329).

Livshits, M. S., 1975, Correlation model for the recognition of objects by echolocating animals, Biofizika, 20(5):920 (English translation JPRS L/5744).

Ljungblad, D. K., Leatherwood, S., and Johnson, R. A., 1977, Echolocation signals of wild Pacific bottlenosed dolphins Tursiops Sp, in: "Proceedings (Abstracts) Second Conference on the Biology of Marine Mammals", San Diego.

Lockyer, C., 1977, Observations on diving behaviour of the sperm whale Physeter catodon, in: "A Voyage of Discovery", M. Angel, ed., Pergamon Press, Oxford.

Love, R. H., 1971, Measurements of fish target strength: a review, Fish. Bull., 69:703.

Love, R. H., 1973, Target strengths of humpback whales, Megaptera novaeangliae, Jour. Acous. Soc. Amer., 54:1312.

Mackay, R. S., 1966, Telemetering physiological information
    from within cetaceans, and the applicability of ultrasound
    to understanding in vivo structure and performance, in:
    "Whales, Dolphins and Porpoises", K. S. Norris, ed.,
    Univ. Calif. Press, Berkeley.

MacKay, R. S., Rumage, W. T., and Becker, A., 1977, Sound veloc-
    ity in spermaceti organ of a young sperm whale, in:
    "Proceedings (Abstracts) Second Conference on the
    Biology of Marine Mammals", San Diego.

Malins, D. C., and Varanasi, V., 1977, Acoustic pathways in
    the cetacean head: Assessment of sound properties
    through the use of a new microtechnique, in: "Proceedings
    (Abstracts) Second Conference on the Biology of Marine
    Mammals", San Diego.

Markov, V. I., and Ostrovskaya, V. M., 1975, On the "identifica-
    tion" signal of dolphins, in: "Morskiye Mlekopitayushchiye",
    G. B. Agarkov, ed., Izdatel'stov Naukova Dumka, Kiev.

Markov, V. I., Tanchevskaya, V. A., and Ostrovskaya, V. M.,
    1974, Organization of acoustic signals in the Black
    Sea bottlenose dolphins (Tursiops truncatus), in:
    "Morfologiya, Fiziologiya i Akustika Morskikh
    Mlekopitayushchikh", V. Ye. Sokolov, ed., Izdatel'stvo
    Nauka, Moscow.

McCormick, J. G., 1968, "Theory of hearing for delphinids",
    Doctoral Dissertation, Princeton University, New
    Jersey.

McCormick, J. G., 1972, The physiology of hearing in the porpoise,
    in: "Marine Mammals: Biology and Medicine", S. H. Ridgway,
    ed., Thomas, Springfield, Ill.

McCormick, J. G., Wever, E. G., Palin, J. and Ridgway, S. H.,
    1970, Sound conduction in the dolphin ear, Jour. Acoust.
    Soc. Amer., 48:1418.

McCormick, J. G., Wever, E. G., Ridgway, S., and Palin, J.,
    1970, Function of the porpoise ear as shown by its
    electrical potentials (Abstract), Jour. Acoust. Soc.
    Amer., 47:67.

McDonald-Renaud, D. L., 1974, Sound localization in the
    bottlenose porpoise, Tursiops truncatus (Montague),
    Doctoral Dissertation, Univ. of Hawaii, Honolulu.

Mead, J. G., 1972, On the anatomy of the external nasal passages
    and facial complex in the family delphinidae of the order
    cetacea, Doctoral Dissertation, Univ. Chicago, Chicago.

Mermoz, H., 1967, Discussion, in: "Animal Sonar Systems, Biology
    and Bionics", R.-G. Busnel, ed., Laboratoire de Physiologie
    Acoustique, Jouy-en-Josas, France.

Metsaveer, J., 1971, Algorithm for calculation of echo-pulses
    from elastic spherical shells in fluid by summation
    of wave groups, Inst. of Cybernetics, Acad. Sci. Eston.
    SSR Tallinn (Preprint), 3:1.

Miyasnikov, V. S., and Titov, M. S., 1975, On the identifi-
    cation of marine animals by their acoustic signals,
    Izvestiya TINRO, 94:38.

Mizue, T., Nishiwaki, M., and Takemura, A., 1971, The underwater
    sound of Ganges river dolphins (Platanista gangetica),
    Sci. Repts. Whales Res. Inst., 23:123.

Mizue, T., Takemura, A., and Nakasai, K., 1967, Studies on
    the little toothed whales in the west sea area of Kyushu,
    XIII, Mating calls and others of the bottlenose dolphin
    caught at Arikawa in Goto Is., Nagasaki Pref. Bull. Fac.
    Fish. Nagasaki Univ., 23:197.

Mizue, T., Takemura, A., Nakasai, K., 1968, Studies on the
    little toothed whales in the west sea area of Kyushu,
    XV, Underwater sound of the Chinese finless porpoise
    caught in the Japanese coastal sea, Bull. Fac. Fish.
    Nagasaki Univ., 25:25.

Mohl, B., 1967, Frequency discrimination in the common seal
    and a discussion of the concept of upper hearing limit,
    in: "Underwater Acoustics", V. M. Albers, ed., Plenum
    Press, New York.

Mohl, B., 1968, Auditory sensitivity of the common seal in
    air and water, Jour. Aud. Res., 8:27.

Mohl, B., 1968, Hearing in seals, in: "The Behavior and
    Physiology of Pinnipeds", R. J. Harrison, R. C. Hubbard,
    R. S. Peterson, C. E. Rice, and R. J. Schusterman, eds.,
    Appleton-Century-Crofts, New York.

Mohl, B., and Andersen, S., 1973, Echolocation: high-frequency
    component in the click of the harbour porpoise (Phocoena
    Ph. L), Jour. Acoust. Soc. Amer, 54:1368.

Mohl, B., Ronald, K., and Terhune, J. M., 1972, The harp seal,
    Pagophilus groenlandicus (Erxleben, 1777), XVIII, Under-
    water calls, in: "Symposium on the Biology of the
    Seal", G. B. Farquhar, ed., Univ. of Guelph, Ontario.

Moore, J. C., 1968, Relationships among the living genera of
    beaked whales, Fieldiana:Zoology, 53(4):209.

Moore, P. 1975, Underwater localization of click and pulsed
    pure tone signals by the California sea lion (Zalophus
    californianus), Jour. Acoust. Soc. Amer., 57:406.

Moore, P., and Au, W., 1975, Underwater localization of pulsed
    pure tones by the California sea lion (Zalophus
    californianus), Jour. Acoust. Soc. Amer., 58:721.

Moore, P. W. B., and Schusterman, R. J., 1977, The upper limit
    of underwater auditory frequency discrimination in the
    California sea lion, in: "Proceedings (Abstracts) Second
    Conference on the Biology of Marine Mammals", San Diego.

Morgan, D. W., 1971, The reactions of belugas to natural
    sound playbacks, in: "Proceedings of the Seventh Annual
    Conference on Biological Sonar and Diving Mammals",
    Stanford Research Institute, Menlo Park, Calif.

Morgane, P. J., and Jacobs, N.S., 1972, Comparative anatomy
    of the cetacean nervous system, in: "Functional Anatomy
    of Marine Mammals", R. J. Harrison, ed., Academic Press,
    New York.

Morii, H., and Kanazu, R., 1972, Fatty acids in the foetus,
    nurseling and adult of a kind of dolphin, Stenella
    attenuata, Bull. Jap. Soc. Scient. Fish., 38:599.

Morozov, V. P., Akopian, A. I., Zaytseva, K. A. and Titov, A. A.,
    1975, On the characteristics of spacial directivity of the
    dolphin acoustic system in signal perception against the
    background of noise, in: "Marine Mammals, Proceedings of the
    Sixth All-Union Conference on the Study of Marine Mammals",
    G. B. Agarkov, ed., Naukova Dumka, Kiev (English translation
    JPRS L/6049).

Morozov, V. P., Akopian, A. I., Burdin, V. I., Zaytseva, K. A.
    and Sokovykh, Yu. A., 1972, Tracking frequency of the loca-
    tion signals of dolphins as a function of distance to the
    target, Biofizika, 17:139 (English translation JPRS 55729).

Morozov, V. P., Akopian, A. I., Burdin, V. I., Donskov, A. A.,
    Zaytseva, K. A., and Sokovykh, Yu. A., 1971, Audiogram
    of the dolphin, Tursiops truncatus, Fiziologicheskiy
    Zhurnal imeni I. M. Sechenova, 57(6):843.

Morris, R. J., 1973, The lipid structure of the spermaceti organ
    of the sperm whale (Physeter catodon), Deep-Sea Res., 20:911.

Murchison, A. E., 1976, Range Resolution by an Echolocating Dolphin
    (Tursiops truncatus), Jour. Acoust. Soc. Amer., 60:S5.

Murchison, A. E., 1979, "Maximum Detection Range and Range
    Resolution in Echolocating Tursiops truncatus (Montague)",
    Doctoral Dissertation U. of Calif. at Santa Cruz, Calif.
    (manuscript).

Murchison, A. E., and Nachtigall, P. E., 1977, Three dimensional
    shape discrimination by an echolocating bottlenose porpoise
    (Tursiops truncatus), in: "Proceedings (Abstracts) of the
    Second Conference on the Biology of Marine Mammals",
    San Diego.

Murchison, A. E., and Penner, R. H., 1975, Open water echolocation
    in the bottlenose dolphin (Tursiops truncatus): Metallic
    sphere detection thresholds as a function of distance, in:
    "Conference on the Biology and Conservation of Marine
    Mammals, Abstracts", 2:42.

Nachtigall, P. E., 1969, Visual size discrimination in the
    East Asian clawless otter (Amblyonyx cineria) in air and
    under water, in: "Proceedings of the Sixth Annual Conference
    on Biological Sonar and Diving Mammals", Stanford Research
    Institute Press, Menlo Park, Calif.

Nachtigall, P. E., 1971, Spatial discrimination and reversal
    based on differential magnitude of reward in the dolphin
    Tursiops truncatus, in: "Proceedings of the Eighth Annual
    Conference on Biological Sonar and Diving Mammals",
    Stanford Research Institute, Menlo Park, Calif.

Nachtigall, P. E., 1976, "Food-Intake and food-rewarded instru-
    mental performance in dolphins as a function of feeding
    schedule", Doctoral Dissertation, Univ. of Hawaii, Honolulu.
Nachtigall, P. E., 1977, A comparison of porpoise (Tursiops
    truncatus) performance with fish and a prepared food,
    in: "Proceedings (Abstracts) Second Conference on the
    Biology of Marine Mammals", San Diego.
Nachtigall, P. E., Murchison, A. E., and Au, W. W. L., 1978,
    Discrimination of solid cylinders and cubes by a blind-
    folded echolocating bottlenose dolphin (Tursiops truncatus),
    Jour. Acoust. Soc. Amer., 64(1):587.
Naumov, N. P., 1973, Signal (biological) fields and their
    significance for animals, Zhurnal Obshchey Biologii, 6:808.
Naumov, N. P., 1975, Biological (signal) fields and their
    significance in mammalian life, Vestnik Akademii Nauk
    SSSR, 45(2):55.
Naumov, N. P., 1975, Bionics and acoustics, Berliner Zeitung,
    (E. Germany) Lomonosov St., Univ., 4.
Neproshin, A. Yu., 1975, Sounds as part of the behavior of the
    Pacific beluga, Priroda, 4:99 (English translation: JPRS
    65658).
Ness, A. R., 1967, A measure of asymmetry of the skulls of odonto-
    cete whales, Jour. Zool., 153:209.
Norris, K. S., 1968, The evolution of acoustic mechanisms in
    odontocete cetaceans, in: "Evolution and Environment", E. T.
    Drake, ed., Yale Univ. Press, New Haven.
Norris, K. S., 1969, The echolocation of marine mammals, in:
    "The Biology of Marine Mammals", H. T. Andersen, ed.,
    Academic Press, New York.
Norris, K. S., 1974, "The Porpoise Watcher", George J. McLeod
    Limited, Toronto.
Norris, K. S. and Evans, W. E., 1967, Directionality of echolocation
    clicks in the rough-tooth porpoise Steno bredanensis (Lesson),
    in: "Marine Bio-Acoustics, Proceedings Second Symposium on
    Marine Bio-Acoustics, New York", W. N. Tavolga, ed., Pergamon
    Press, New York.
Norris, K. S., and Harvey, G. W., 1972, A theory for the function
    of the spermaceti organ of the sperm whale, in: "Animal
    Orientation and Navigation, S. R. Galler et al., eds., National
    Aeronautics and Space Administration, Washington, D. C.
Norris, K. S., and Harvey, G. W., 1974, Sound transmission in
    the porpoise head, Jour. Acoust. Soc. Amer., 56:659.
Norris, K. S., and Watkins, W. A., 1971, Underwater sounds of
    Arctocephalus philippii, the Juan Fernandez fur seal,
    Antarct. Res. Ser., 18:169.
Norris, K. S., Evans, W. E. and Turner, R. N., 1967, Echolocation
    of an Atlantic bottlenose porpoise during discrimination,
    in: "Animal Sonar Systems, Biology and Bionics", R.-G.
    Busnel, ed., Laboratoire de Physiologie Acoustique, Jouy-en-
    Josas, France.

Norris, K. S., Dormer, K. J., Pegg, J., and Liese, G. J., 1971,
    The mechanisms of sound production and air recyling in
    porpoises: A preliminary report, in: "Proc. 8th Annual
    Conf. Biol. Sonar and Diving Mammals", Stanford Research
    Institute, Menlo Park, Calif.

Norris, K. S., Harvey, G. W., Burzell, L. A., and Krishna Kartha,
    T. D., 1972, Sound production in the freshwater porpoises
    Sotalia and Inia, in Rio Negro, Brazil, in: "Investigations
    on Cetacea Vol. 4., G. Pilleri, ed., Benetelli AG, Berne.

Northrop, J., Cummings, W. C., and Morrison, M. F., 1971,
    Underwater 20-Hz signals recorded near Midway Island,
    Jour. Acoust. Soc. Amer., 49:1909.

Northrop, J. , Cummings, W. C., and Thompson, P. O., 1968, 20-Hz
    signals observed in the central pacific, Jour. Acoust. Soc.
    Amer., 43:383.

Ostroumov, G. A., 1968, At what distances could marine animals
    communicate by means of electromagnetic waves (in problems
    of wave propagation), Izdatel'stvo Leningradskogo
    Universiteta, 8:3.

Paul, L. J., 1969, Dolphin noises recorded by echosounder, New
    Zealand Jour. Marine and Freshwater Res., 3:343.

Payne, R. S., and McVay, S., 1971, Songs of humpback whales,
    Science, 173:587.

Payne, R. S., and Payne, K., 1972, Underwater sounds of southern
    right whales, Zoologica, 56:159.

Payne, R., and Webb, D., Orientation by means of long-range
    acoustic signaling in Baleen whales, in: "Orientation:
    Sensory Basis", H. E. Adler, ed., Ann. N. Y. Acad. Sci.,
    188:110.

Penner, R. H., 1977, Paired simultaneous echo ranging by Tursiops
    truncatus, in: "Proc. (Abstracts) Second Conference on the
    Biology of Marine Mammals", San Diego.

Penner, R., and Murchison, A. E., 1970, Experimentally demonstrated
    echolocation in the Amazon river porpoise Inia geoffrensis
    (Blainville), in: "Proceedings of the Seventh Annual
    Conference on Biological Sonar and Diving Mammals", Stanford
    Research Institute, Menlo Park, Calif.

Penner, R. H., and Murchison, A. E., 1970, Experimentally
    demonstrated echolocation in the Amazon River porpoise,
    Inia geoffrensis (Blainville), Naval Undersea Center,
    Technical Publication 187, San Diego, Calif.

Perkins, P. J., 1966, Communication sounds of finback whales,
    Norsk Hvalfangst-Tidend, 55:199.

Perkins, P. J., Fish, M. P., and Mowbray, W. H., 1966, Underwater
    communication sounds of sperm whale Physeter catodon,
    Norsk Hvalfangst-Tidende, 55:225.

Perrin, W. F., and Hunter, J. R., 1972, Escape behavior of the
    Hawaiian spinner porpoise (Stenella cf. S. longirostris), U.
    S. Nat. Mar. Fish. Bull., 70:49.

Perryman, W. and Au, D., 1977, Aerial observations of evasive
    behavior of dolphin schools, in: "Proceedings 2nd Conference
    Bio. Mar. Mammals", San Diego.

Pilleri, G., 1970, Records of cetaceans off the Italian and
    Dalmatian coasts, in: "Investigations on Cetacea, Vol. 2",
    G. Pilleri, ed., Bentelli AG, Berne.

Pilleri, G., and Busnel, R.-G., 1969, Brain/body weight ratios
    in Delphinidae, Acta. Anat. 73:92.

Pilleri, G., and Gihr, M., 1970, Brain-body weight ratio of
    Platanista gangetica, in: "Investigations on Cetacea, Vol.
    2", G. Pilleri, ed., Bentelli AG, Berne.

Pilleri, G., Kraus, C., and Gihr, M., 1970, Frequenzanalyse
    der Laute von Platanista indi (Cetacea), Rev. Suisse
    Zool., 77:922.

Pilleri, G., Kraus, C., and Gihr, M., 1971, Physical analysis
    of the sounds emitted by Platanista indi, in: "Investigation
    on Cetacea, Vol. 3", G. Pilleri, ed., Bentelli AG, Berne.

Popov, V. V., and Supin, A. Yu., 1976, Determination of the
    hearing characteristics of dolphins by measuring induced
    potentials, Fiziol. Zh. SSSR im I. M. Sechenova, 62:550.

Poulter, T. C., 1966, Biosonar, in: "McGraw-Hill Yearbook
    of Science and Technology", McGraw-Hill, New York.

Poulter, T. C., 1966, The use of active sonar by the California
    sea lion Zalophus californianus (Lesson), Jour. Aud. Res.,
    6:165.

Poulter, T. C., 1967, Systems of echolocation, in: "Animal
    Sonar Systems: Biology and Bionics", R.-G. Busnel, ed.,
    Laboratoire de Physiologie Acoustique, Jouy-en-Josas,
    France.

Poulter, T. C., 1968, Marine mammals, in: "Animal Communication",
    T. A. Sebeok, ed., Indiana University Press, Bloomington.

Poulter, T. C., 1968, Underwater vocalization and behavior of
    pinnipeds, in: "The Behavior and Physiology of Pinnipeds",
    R. J. Harrison, et. al., eds., Appleton-Century-Crofts,
    New York.

Poulter, T. C., 1968, Vocalization of the gray whales in Laguna
    Oja de Liebre (Scammon's Lagoon), Baja, California,
    Mexico, Norsk Hvalfangsttid, 57:53.

Poutler, T. C., 1969, Conditioning marine mammals for testing
    echolocation abilities, in: "Proceedings of the Sixth
    Annual Conference on Biological Sonar and Diving Mammals",
    Stanford Research Institute, Menlo Park, Calif.

Poulter, T. C., 1969, Sonar discrimination ability of the
    California sea lion, Proceedings California Acad. Sci.
    36:381.

Poulter, T. C., 1969, Sonar of penguins and fur seals, Proceedings
    California Academy Science ", 36:363.

Poulter, T. C., 1970, Ultrasonic frequencies recorded from three
    captive blind dolphins, Platanista gangetica, in: "Proceed-
    ings of the Seventh Annual Conference on Biological
    Sonar and Diving Mammals", Stanford Research Institute,
    Menlo Park, Calif.

Poulter, T. C., 1972, Sea lion vibrissae - an acoustic sensor,
    in: "Proceedings of the Ninth Annual Conference on Biologi-
    cal Sonar and Diving Mammals", Stanford Research Institute,
    Menlo Park, Calif.

Powell, B. A., 1966, Periodicity of vocal activity of captive
    Atlantic bottlenose dolphins, Tursiops truncatus,
    Bull. So. Calif. Acad. Sci., 65:237.

Pryor, K. and Norris, K. S., 1978, The tuna porpoise problem:
    Behavioral aspects, Oceanus, 21:31.

Purves, P. E., 1966, Anatomy and Physiology of the outer
    middle ear in cetaceans, in: "Whales, Dolphins and
    Porpoises", K. S. Norris, ed., Univ. Calif. Press,
    Berkeley.

Purves, P. E., 1967, Anatomical and experimental observations
    on the cetacean sonar systems, in: "Animal Sonar Systems,
    Biology and Bionics", R.-G. Busnel, ed., Laboratoire de
    Physiologie Acoustique, Jouy-en-Josas, France.

Purves, P. E., and Pilleri, G., 1973, Observations on the ear,
    nose, throat and eye of Platanista indi., in: "Investi-
    gations on Cetacea", G. Pilleri, ed., Bentelli AG, Berne.

Ray, C., 1967, Social behavior and acoustics of the Weddell seal,
    Antarct. Jour. U. S., 2(4):105.

Ray, G. C., 1972, "Marine mammals: The biological basis of produc-
    tivity and conservation", Smithsonian Inst., Wash. D. C.

Ray, G. C., and Watkins, W. A., 1972, Social function of phonation
    in walrus, in: "Symposium on the Biology of the Seal",
    G. B. Farquhar, ed., Univ. of Guelph, Ontario.

Ray, G. C., Watkins, W. A., and Burns, J. J., 1969, The underwater
    song of Erignathus (bearded seal), Zoologica 54:79.

Renaud, D. L., and Popper, A. N., 1975, Sound localization by the
    bottlenose porpoise, Tursiops truncatus, Jour. Exp. Biol.
    63:569.

Repenning, C. A., Underwater hearing in seals: Functional
    morphology, in: "Functional Anatomy of Marine Mammals",
    R. J. Harrison, ed., Academic Press, New York.

Resvov, R. N., Savel'ev, Ya. A., and Flenov, A. B., 1973, The
    study of dolphin's echolocator ability in detection of
    different targets, difference in range and angle on vertical
    and horizontal planes, in: "Proceedings of the 8th All-Union
    Acoustical Conference", Leningrad.

Reysenbach de Haan, F. W., 1966, Listening underwater: Thoughts
    on sound cetacean hearing, in: "Whales, Dolphins, and
    Porpoises", K. S. Norris, ed., Univ. of Calif. Press,
    Berkeley.

Reznik, A. M., Skornyakov, V. M., and Chupakov, A. G., 1970,
    Echolocation activity of Black Sea Tursiops truncatus
    being presented targets, Trudy Akusticheskogo Instituta,
    12:116.

Reznik, A. M. Skornyakov, V. M., Chernyshev, O. B., and
    Chupakov, A. G., 1968, On some peculiarities of dolphin
    communication system, in: "Trudy VIoy Vsesoijuznoyi
    Akusticheskoi Konferentsii", Moscow.

Reznikov, A. E., 1970, On sound-vision in dolphins, in: "Tezisy
    Dokladov 23ey Nauchno-tekhnicheskoy. Konferentsii Lenin
    gradskogo Institute Aviatsionnogo Priborostroeyniya",
    Leningrad.

Reznikov, A. E., 1971, Mechanisms of delphinid hydroacoustic
    ranging, Priroda, 9:171.

Rice, D. W., 1978, Beaked whales, in: "Marine Mammals", D.
    Haley, ed., Pacific Search Press.

Ridgway, S. H., 1966, Dall porpoise, Phocoenoides dalli (True):
    Observations in captivity and at sea, Norsk Hvalfangst-
    Tidende, 5:97.

Ridgway, S. H., 1972, (ed.) "Mammals of the Sea, Biology and
    Medicine", Charles C. Thomas, Springfield, Ill.

Ridgway, S. H., and Brownson, R. W., 1979, Relative brain
    sizes and cortical surface areas in odontocetes,
    (In preparation).

Ridgway, S. H., and McCormick, J. G., 1967, Anesthetization
    of porpoises for major surgery, Science, 158:510.

Ridgway, S. H., Flanagan, N. J., and McCormick, J. G., 1976,
    Brain-spinal cord ratios in porpoises: possible correlation
    with intelligence and ecology, Psychon. Sci., 6:491.

Ridgway, S. H., McCormick, J. G., and Wever, E. G., 1974, Surgical
    approach to dolphin's ear, Jour. Exp. Zool., 188:265.

Ridgway, S. H., Scronce, B. L., and Kanwisher, J., 1969,
    Respiration and deep diving in the bottlenosed porpoise,
    Science, 166:1651.

Ridgway, S. H., Carder, D. A., Green, R. F., Gaunt, A. S.,
    Gaunt, S. L., and Evans, W. E., 1977, Electromyography
    and dolphin sound production, in: "Proceedings (Abstracts)
    Second Conference on the Biology of Marine Mammals, San
    Diego.

Rieger, M. F. P., and Johnson, R. A., 1977, Chorusing patterns
    of pilot whales, Globicephala macrorynchus, in: "Proceedings
    (Abstracts) Second Conference on the Biology of Marine
    Mammals", San Diego.

Robisch, P. A., Malins, D. C., Best, R., and Varanasi, V., 1972,
    Differences in triacylglycerols from acoustic tissues and
    posterior cranial blubber of the narwhal (Monodon monoceros),
    Biochem. Jour., 130:33.

Rodionov, V. A., 1975, Morpho-functional features of the
    respiratory musculature of dolphins, Zoologicheskiy
    Zhurnal, 53(6):919.

Romanenko, Ye. V., 1972, Near acoustic field of the bottlenose dolphin (Tursiops truncatus), Voe Vsesoyuznoye Soveshchaniye po izucheniyu morskikh mlekopitayushchikh, Makhachkala, Part II:200.

Romanenko, Ye. V., 1973, Mechanism of emission and formation of echolocating signals of the bottlenose dolphin (Tursiops truncatus), in: "Eighth Annual All-Union Acoustical Conference", Moscow.

Romanenko, Ye. V., 1973, Sound signals in the near field of the Atlantic bottlenose dolphin (Tursiops truncatus, Delphinidae), Zoologicheskiy Zhurnal, 52(11):1698 (English translation JPRS 61553).

Romanenko, Ye. V., 1974, On the mechanism of the dolphin emission of some pulse and whistle signals, in: "Morfologiya, Fiziologiya i Akustika Morskikh Mlekopitayushchikh", V. Ye. Sokolov, ed., Izdatel'stvo Nauka, Moscow.

Romanenko, Ye. V., 1974, "Physical fundamentals of bioacoustics", Fizichekiye Osnovy Bioakustik, Moscow (English translation JPRS 63923).

Romanenko, Ye. V., 1975, On the hearing of dolphins, in: "Morskiye Mlekopitayushchiye", G. B. Agarkov, ed., Izdatel'stvo Naukova Dumka, Kiev.

Romanenko, Ye. V., 1975, Recording sounds in the dolphin's respiratory system, in: "Morskiye Mlekopitayushchiye", G. B. Agarkov, ed., Izdatel'stvo Naukova Dumka, Kiev.

Romanenko, Ye. V., Tomilin, A. G. and Artemenko, B. A., 1965, in: "Bionics", M. G. Gaaze-Rapoport and V. E. Yakobi, eds., Nauka, Moscow (English translation JPRS 35125).

Romanenko, Ye., V., Yanov, V. G., and Akopian, A. I., 1974, A method of studying the echolocating system of the dolphin using a radiotelemetry system, in: "Morfologiya, Fiziologiya i Akustika Morskikh Mlekopitayushchikh", V. Ye. Sokolov, ed., Izdatel'stvo Nauka, Moscow.

Ryan, W. W., Jr., 1978, Acoustical reflections from aluminum cylindrical shells immersed in water, Jour. Acous. Soc. Amer., 64:1159.

Sales, G., and Pye, D., 1974, "Ultrasonic Communication by Animals," Chapman and Hall, London.

Saprykin, V. A., Korolev, V. I., Kovtuneko, S. V., and Dmitriyeva, Ye. S., 1973, An investigation of spatial frequency characteristics of the Tursiops echolocator in object identification, Doklady Akademii Nauk SSSR, 214(3):727.

Saprykin, V. A., Korolev, V. I., Kovtunenko, S. V., Dmitriyeva, Ye. S., and Sokovykh, Yu. A., 1975, Spatial analysis of the tonal supra threshold signals in dolphins, Biofizika, 20:319.

Saprykin, V. A., Kovtuneko, S. V., Ekker, I. V., Dmitriyeva, Ye. S., and Korolev, V. I., 1975, Invariance of the sensing properties of the auditory analyser of the dolphin in differentiating pulse tone signal, Doklady Akademii Nauk SSSR, 221(4):999.

Saprykin, V. A., Kovtuneko, S. V., Korolev, V. I., Dmitriyeva,
    Ye. S., and Belov, B. I., 1975, Study of the auditory percep-
    tion of dolphins as a function of signal characteristics in
    the time domain, Biofizika, 20(4):720.
Saprykin, V. A., Kovtuneko, S. V., Korolev, V. I., Dmitriyeva,
    Ye. S., Belov, B. I., and Mar'yasin, V. G., 1975, Study of the
    auditory perception of dolphins as a function of signal char-
    acteristics in the time domain, Biofizika, 20 (English
    translation JPRS 65877).
Saprykin, V. A., Kovtuneko, S. V., Korolev, V. I., Dmitriyeva,
    Ye. S., Ol'shanskiy, V. I., and Bekker, I. V., 1976, Invari-
    ance of the auditory perception relative to frequency-time
    signal transformations in the dolphin Tursiops truncatus,
    Zhurnal Evolyutsionnoy Biokhimii I Fiziologii (English
    translation JPRS L/6275).
Schenkkan, E. J., 1971, The occurrence and position of the "con-
    necting sac" in the nasal tract complex of small odontocetes
    (Mammalia, Cetacea), Beaufortia, 19:37.
Schenkkan, E. J., 1972, On the nasal tract complex of Pontoporia
    blainvillei GERVAIS and d'ORBIGNY, 1884 (Cetacea, Plata-
    nistidae), in: "Investigations on Cetacea Vol 4",
    G. Pilleri, ed., Bentelli AG, Berne.
Schenkkan, E. J., 1973, On the comparative anatomy and function of
    the nasal tract in odontocetes (Mammalia, Cetacea), Bijdragen
    Tot de Dierkunde, 43(3):127.
Schenkkan, E. J., and Purves, P. E., 1973, The comparative anatomy
    of the nasal tract and the function of the spermaceti organ
    in the Physeteridae (Mammalia, Odontoceti), Bijdragen Tot
    de Dierkunde, 43(1):93.
Schevill, W. E., 1968, Sea lion echo ranging?, Jour. Acoust. Soc.
    Am., 43:1458.
Schevill, W. E., and Watkins, W. A., 1966, Sound structure and
    directionality in Orcinus (Killer Whale), Zoologica, 51:71.
Schevill, W. E., and Watkins, W. A., 1971, Directionality of the
    sound beam in Leptonychotes weddelli (Mammalia, Pinnipedia),
    Antarct. Res. Ser., 18:163.
Schevill, W. E., and Watkins, W. A., 1971, Pulsed sounds of the
    porpoise Lagenorhynchus australis, Breviora, 366:1.
Schevill, W. E., and Watkins, W. A., 1972, Intense low-frequency
    sounds from an Antarctic minke whale, Balaenoptera acutoro-
    strata, Breviora, 388:1.
Schevill, W. E., Watkins, W. A., and Ray, G. C., 1966, Analysis of
    underwater Odobenus calls with remarks on the development and
    function of the pharyngeal pouches, Zoologica, 51:103.
Schevill, W. E., Watkins, W. A., and Ray, G. C., 1969, Click struc-
    ture in the porpoise, Phocoena phocoena, Jour. Mamm., 50:721.
Schusterman, R. J., 1966, Serial discrimination-reversal learning
    with and without errors by the California sea lion, J. Exp.
    Analysis Behav., 9(5):97.

Schusterman, R. J., 1966, Underwater click vocalizations by a
    California sea lion: effects of visibility, Psychol. Rec.,
    16:129.
Schusterman, R. J., 1967, Perception and determinants of under-
    water vocalization in the California sea lion, in: "Animal
    Sonar Systems, Biology and Bionics", R.-G. Busnel, ed.,
    Laboratoire de Physiologie Acoustique, Jouy-en-Josas, France.
Schusterman, R. J., 1968, Experimental laboratory studies of pin-
    niped behaviour, in: "The Behaviour and Physiology of Pin-
    nipeds", R. J. Harrison, R. C. Hubbard, R. S. Peterson, C. E.
    Rice and R. J. Schusterman, eds., Appleton-Century-Crofts,
    New York.
Schusterman, R. J., 1969, Aerial and underwater visual acuity in
    the California sea lion as a function of luminance, Naval
    Undersea Research and Development Center, San Diego, Calif.,
    Final Report.
Schusterman, R. J., 1972, Visual acuity in pinnipeds, in: "Beha-
    vior of Marine Animals Vol. II", H. E. Winn and B. Olla,
    eds., Plenum Press, New York.
Schusterman, R. J., 1973, A note comparing the visual acuity of
    dolphins with that of sea lions, Cetology, 15:1.
Schusterman, R. J., 1974, Low false-alarm rates in signal detec-
    tion by marine mammals, Jour. Acoust. Soc. Amer., 55:845.
Schusterman, R. J., 1976, California sea lion auditory detection
    and variation of reinforcement schedules, Jour. Acoust. Soc.
    Amer., 59:997.
Schusterman, R. J., 1977, Temporal patterning in sea lion barking
    (Zalophus californianus), Behavioral Biology, 20:404.
Schusterman, R. J., and Balliet, R. F., 1969, Underwater barking by
    male sea lions (Zalophus californianus), Nature 222:1179.
Schusterman, R. J., and Balliet, R. F., 1970, Conditioned vocaliza-
    tion as a technique for determining visual acuity thresholds
    in the sea lion, Science, 169:498.
Schusterman, R. J., and Balliet, R. F, 1970, Visual acuity of the
    harbor seal and the Steller sea lion under water, Nature,
    226:563.
Schusterman, R. J., and Balliet, R. F., 1971, Aerial and underwater
    visual acuity in the California sea lion (Zalophus californi-
    anus) as a function of luminance, Ann. N.Y. Acad. Sci.,
    188:37.
Schusterman, R. J., and Barrett, B., 1973, Amphibious nature of
    visual acuity in the Asian 'clawless' otter, Nature, 244:518.
Schusterman, R. J., and Dawson, R. G., 1968, Barking, dominance,
    and territoriality in male sea lions, Science, 160:434.
Schusterman, R. J., and Johnson, B. W., 1975, Signal probability
    and response bias in California sea lions, Psych. Rec.,
    25:39.

Schusterman, R. J., and Kersting, D., 1978, Selective attention in discriminative echolocation by the porpoise (Tursiops truncatus), Paper read at the Animal Behavior Society Annual Meeting, June 19-23, U. of Wash., Seattle.

Schusterman, R. J., Balliet, R. F., and Nixon, J., 1972, Underwater audiogram of the California sea lion by the conditioned vocalization technique, J. exp. Analysis Behav., 17:339.

Schusterman, R. J., Balliet, R. F., and St. John, S., 1970, Vocal displays by the gray seal, the harbor seal and the Steller sea lion, Psychon. Sci., 18:303.

Schusterman, R. J., Barrett, B., and Moore, P. W. B., 1975, Detection of underwater signals by a California sea lion and a bottlenose porpoise: Variation in the payoff matrix, Jour. Acoust. Soc. Amer., 57:1526.

Scronce, B. L., and Johnson, C. S., 1975, Bistatic target detection by a bottlenosed porpoise, Jour. Acoust. Soc. Am., 59(4):1001.

Seeley, R. L., Flanigan, W. F., and Ridgway, S. H., 1976, A technique for rapidly assessing the hearing of the bottlenosed porpoise Tursiops truncatus, Naval Undersea Center TP 522, San Diego, Calif.

Sergeant, D. E., 1978, Ecological isolation in some cetacea, in: "Recent Advances in the Study of Whales and Seals", A. N. Severtsov, ed., Nauka, Moskow.

Shevalev, A. Ye,. and Flerov, A. I., 1975, Study of the communicative signal of dolphins connected with food-finding, Bionika, 9:119.

Shirley, D. J., and Diercks, K. J., 1970, Analysis of the frequency response of simple geometric targets, Jour. Acoust. Soc. Am., 48:1275.

Sokolov, V., 1971, Cetacean research in the USSR, in: "Investigations on Cetacea Vol. 3", G. Pilleri, ed., Bentelli AG, Berne.

Sokolov, V., and Kalashnikova, M., 1971, The ultrastructure of epidermal cells in Phocoena phocoena, in: "Investigations on Cetacea Vol. 3", G. Pilleri, ed., Bentelli AG, Berne.

Sokolov, V. Ye., Ladygina, T. F., and Supin, A. Ya., 1972, Localization of sensory areas in the dolphin's cerebral cortex, Doklady Akademii Nauk SSSR, 200(2).

Solntseva, G. N., 1971, The comparative anatomical and histological characteristics of the structure of the external and internal ear of certain dolphins, Issledovaniya Morskikh Mlekopitayushchikh, Trudy Atlantnipo, 39:369.

Solntseva, G. N., 1974, Function of the hearing organ underwater, Priroda, 10:21 (English translation JPRS 65061).

Solntseva, G. N., 1975, Objectives and prospects of research on routes of sound transmission in cetaceans, in: "Morskiye Mlekopitayushchiye", G. B. Agarkov, ed., Izdatel'stvo Naukova Dumka, Kiev.

Solntseva, G. N., 1975, Morpho-functional peculiarities of the hearing organ in terrestrial, semiaquatic and aquatic mammals, Zoologicheskiy Zhurnal, 54(10):1529.

Solukha, B. V., and Mantulo, A. P., 1973, Mechanism of sound transmission in dolphins, in: "Nekotorye voprosy ekologii i morfologii zhivotnykh" (Materially Nauchnoy Konferentsii), Izdatel'stvo Naukova Dumka, Kiev.

Stewart, J. L., 1968, Analog simulation studies in echoranging, T. D. R. No. AMRL-TR-68-40, Aerospace Medical Div., U.S.A.F. Systems Command, Washington, D.C.

Sokolov, V. Ye. (ed.), 1978, Electrophysiological studies of the dolphin's brain, Izdatel'stvo, Nauka, Moscow.

Spong, P., and White, D., 1971, Visual acuity and discrimination learning in the dolphin (Lagenorhynchus obliquidens), Exp. Neurol., 31:341.

Spong, P., Bradford, J., and White, D., 1970, Field studies of the behaviour of the killer whale (Orcinus orca), in: "Proceedings of the Seventh Annual Conference on Biological Sonar and Diving Mammals", Stanford Research Institute, Menlo Park, Calif.

Spong, P., Michaels, H., and Spong. L., 1972, Field studies of the behaviour of the killer whale (Orcinus orca) II, in: "Proceedings of the Ninth Annual Conference on Biological Sonar and Diving Mammals", Stanford Research Institute, Menlo Park, Calif.

Spong, P., Spong, L., and Spong, Y., 1972, Field studies of the behaviour of the killer whale (Orcinus orca) III, in: "Proceedings of the Ninth Annual Conference on Biological Sonar and Diving Mammals", Stanford Research Institute, Menlo Park, Calif.

Sukhoruchenko, M. N., 1971, Upper limit of hearing of dolphins with reference to frequency, Trudy Akusticheskogo Instituta, 12:194.

Sukhoruchenko, M. N., 1973, Frequency discrimination of dolphin (Phocoena phocoena), Fiziologicheskiy zhurnal imeni I. M. Sechenova, 59(8):1205.

Supin, A. Ya., and Sukhoruchenko, M. N., 1970, The determination of auditory thresholds in Phocoena phocoena by the method of skin-galvanic reaction, Trudy Akusticheskogo Instituta, 12:194.

Supin, A. Ya., and Sukhoruchenko, M. N., 1974, Characteristics of acoustic analyzer of the Phocoena phocoena L. dolphin, in: "Morphology, Physiology, and Acoustics of Marine Mammals", V. Ye. Sokolov, ed., Nauka, Moscow (English translation JPRS 65139).

Tavolga, W. N., 1977, Mechanisms for directional hearing in the sea catfish (Arius felis), Jour. Exp. Biol., 67:97.

Terhune, J. M, 1974, Directional hearing of a harbor seal in air and water, Jour. Acoust. Soc. Amer., 56:1862.

Terhune, J. M., and Ronald, K., 1972, The harp seal, Pagophilus
    groenlandicus (Erxleben, 1777), X. The underwater audiogram,
    Can. J. Zool., 50:565.
Terhune, J. M., and Ronald, K., 1975, Underwater hearing sensiti-
    vity of two ringed seals (Pusa hispida), Can. J. Zool.,
    53:227.
Terhune, J. M., and Ronald, K., 1975, Masked hearing thresholds of
    ringed seals, Jour. Acoust. Soc., Amer., 58:515.
Thompson, P. O., and Cummings, W. C., 1969, Sound production of the
    finback whale, Balaeoptera physalus, and Eden's whale, B.
    edeni, in the Gulf of California (abstract), in: "Proceed-
    ings of the Sixth Annual Conference on Biological Sonar and
    Diving Mammals", Stanford Research Institute, Menlo Park,
    Calif.
Thompson, R. K. R., and Herman, L. M., 1975, Underwater Frequency
    Discrimination in the Bottlenose Dolphin (1-140 kHz) and
    Human (1-8 kHz), Jour. Acoust. Soc. Amer., 57:943.
Titov, A. A., 1971, Characteristics of sonic signaling of common
    dolphins (Delphinus delphis) under new conditions, Bionika,
    5:62.
Titov, A. A., 1972, Investigation of sonic activity and phenomeno-
    logical characteristics of the echolocation analyzer of Black
    Sea delphinids. Canditorial dissertation, Karadag, as cited
    in: Bel'kovich, V. M. and Dubrovskiy, N. A., 1976, "Sensory
    Bases of Cetacean Orientation", Nauka, Leningrad (English
    translation JPRS L/7157).
Titov, A. A., 1975, Recognition of spherical targets by the bottle-
    nose dolphin in the presence of sonic interference, in:
    "Marine Mammals, Proceedings of the Sixth All-Union Confer-
    ence on the Study of Marine Mammals", G. B. Agarkov, ed.,
    Naukova Dumka, Kiev (English translation JPRS L/6049).
Titov, A. A., and Nikolenko, G. V., 1975, Quantitative evaluation
    of the sounds in three species of Black Sea dolphins,
    Bionika, 9:115.
Titov, A. A., and Tomilin, A. G., 1970, Sonic activity of the
    common dolphin (Delphinus delphis) and harbor porpoise
    (Phocoena phocoena) in various situations, Bionika, 4:88.
Titov, A. A., and Yurkevich, L. I., 1971, Physical characteristics
    of nonecholocation sounds of Black Sea dolphins, Bionika,
    5:57.
Titov, A. A., and Yurkevich, L. I., 1975, Temporary pulse summation
    in the bottlenose dolphin, Tursiops truncatus, in: "Morskiye
    Mlekopitayushchiye", G. B. Agarkov, ed., Izdatel'stvo Naukova
    Dumka, Kiev.
Titov, A. A., Tomilin, A. G., Baryshnikov, N. S., Yurkevich, L. I.,
    and Lekomtsev, V. M., 1971, Communication-emotional signals
    of Black Sea dolphins, Bionika, 5:67.

Tomilin, A. G., 1968, Factors promoting powerful development of
    the brain in Odontoceti, Trudy Vsesoyuznogo Sel'sko-
    Khozyaystvennogo Instituta Zaochnogo Obrasovaniya, 31:191
    (in Russian JPRS 49777).

Tomilin, A. G., 1969, Present status of study of the acoustical
    vocalizations in Cetacea, "IVoe Vsesoyuznoye Soveshchaniye
    po izucheniyu Morskikh Mlekopitayushchikh", Kaliningrad.

Tsuyuki, H., and Itoh, S., 1973, Fatty acid component of blubber
    oil of Amazon river dolphin, Sci. Rept. of the Whales Res.
    Inst., 25:293.

Turner, R. N.,, and Norris, K. S., 1966, Discriminative echolocation
    in a porpoise, Jour. of the Exp. Anal. of Beh., 9:535.

Valiulina, F. G., 1975, The role of the mandible in conduction of
    sonic oscillations in man and dolphins, in: "Morskiye
    Mlekopitayushchiye", G. B. Agarkov, ed., Izdatel'stvo Naukova
    Dumka, Kiev.

Varanasi, V., and Malins, D. G., 1970, Unusual wax esters from the
    mandibular canal of the porpoise (Tursiops gilli),
    Biochemistry, 9:3629.

Varansai, V., and Malins, D. G., 1970, Ester and ether-linked lipids
    in the mandibular canal of a porpoise (Phocoena phocoena).
    Occurrence of isonaleric acid in glycerolipids, Biochemistry,
    9:4576.

Varanasi, V., and Malins, D. G., 1971, Unique lipids of the por-
    poise (Tursiops gilli): difference in triacyl glycerols and
    wax esters of acoustic (mandibular canal and melon) and
    blubber tissues, Biochem. Biophys. Acta., 231:415.

Varanasi, V., and Malins, D. G., 1972, Triacylglycerols character-
    istics of porpoise acoustic tissues: Molecular structure of
    diisolvaleroyl-glycerides, Science, 176:926.

Varanasi, V., Everitt, M., and Malins, D. G., 1973, The isomeric
    composition of the diisovaleroyl-glycerides: a specificity
    for the biosynthesis of the 1,3-Diisolvaleryl-Glycerides
    structures, Int. J. Biochem., 4:373.

Varanasi, V., Feldman, H. R., and Malins, D. G., 1975, Molecular
    basis for formation of lipid sound lens in echolocating
    cetaceans, Nature, 255:340.

Vel'min, V. A., 1975, Target detection by the bottlenose dolphin
    under artifical reverberation conditions, in: "Marine
    Mammals, Proceedings of the Sixth All-Union Conference on
    the Study of Marine Mammals", G. B. Agarkov, ed., Naukova
    Dumka Kiev, (English translation JPRS L/6049).

Vel'min, V. A., and Dubrovskiy, N. A., 1975, On the auditory
    analysis of pulsed sounds by dolphins, Doklady Akad. Nauk.
    SSSR, 225(2):470.

Vel'min, V. A., and Dubrovskiy, N. A., 1976, The critical interval
    of active hearing in dolphins, Sov. Phys. Acoust., 22(4):351,
    (English translation, J. S. Wood, Amer. Inst. Physics).

Vel'min, V. A., and Titov, A. A., 1975, Auditory discrimination of interpulse intervals by bottlenose dolphin, in: "Marine Mammals, Proceedings of the Sixth All-Union Conference on the Study of Marine Mammals", G. B. Agarkov, Ed., Naukova Dumka, Kiev (English translation JPRS L/6049).

Vel'min, V. A., Titov, A. A., and Yurkevich, L. I., 1975, Differential intensity thresholds for short-pulsed signals in bottlenose dolphins, in: "Morskiye Mlekopitayushchiye", (Proceedings of the Sixth All-Union Conference on the Study of Marine Mammals, G. B. Agarkov, ed., Izdatel'stvo Naukova Dumka, Kiev (Translation JPRS L/6049-1).

Vel'min V. A., Titov, A. A., and Yurkevich, L. I., 1975, Temporary pulse summation in bottlenose dolphins, in: "Morskiye Mlekopitayushchiye", (Proceedings of the Sixth All-Union Conference on the Study of Marine Mammals), G. B. Agarkov, ed., Naukova Dumka, Kiev (English Translation JPRS L/6049-1).

Voronov, V. A., and Stosman, I. M., 1977, Frequency-threshold characteristics of subcortical elements of the auditory analyzer of the Phocoena phocoena porpoise, Zhurnal Evolyutsionnoy Biokhimii I Fiziologii, 6:719.

Vronskiy, A. A., 1975, Some data on extraorgan innervation of dolphin pharynx, Doklady Akademii Nauk SSSR, 6:545.

Vronskiy, A. A., and Manger, A. P., 1975, Extraorganic innervation of the larynx and pharynx of some delphinids, in: "Morskiye Mlekopitayushchiye", G. B. Agarkov, ed., Izdatel'stvo Naukova Dumka, Kiev.

Watkins, W. A., 1966, Listening to Cetaceans, in: "Whales, Dolphins and Porpoises", K. S. Norris, ed., Univ. of Calif. Press, Berkeley.

Watkins, W. A., 1967, Air-borne sounds of the humpback whale Megaptera novaengliae, Jour. Mammal, 48:573.

Watkins, W. A., 1967, The harmonic interval: fact or artifact in spectral analysis of pulse trains, in: "Marine Bio-Acoustics Vol. II", W. N. Tavolga, ed., Pergamon Press, New York.

Watkins, W. A., 1968, Comments on "Spectral Analysis of the calls of the Male Killer Whale", IEEE Transact. on Audio and Electroacoustics, 16:523.

Watkins, W. A., 1974, Computer measurement of biological sound-source locations from four hydrophone array data, Technical Report, Off. Naval Res., Washington, D.C.

Watkins, W. A., 1974, Bandwidth limitations and analysis of cetacean sounds, with comments on "Delphinid sonar measurements and analysis", (K. J. Diercks, R. T. Trochta and W. E. Evans, J. Acoust. Soc. Amer. 54, 200-204 (1973)), Jour. Acoust. Soc. Amer., 55:849.

Watkins, W. A., 1976, Biological sound-source location by computer analysis of underwater array data, Deep-Sea Research, 23:175.

Watkins, W. A., 1977, Acoustic behavior of sperm whales, Oceanus, 20:50.

Watkins, W. A., and Ray, G. C., 1977, Underwater sounds from ribbon
    seal, Phoca (Histriophoca) fasciata, Fishery Bulletin, 75:450.
Watkins, W. A., and Schevill, W. E., 1968, Underwater playback of
    their own sounds to Leptonychotes (Weddell seals), Jour.
    Mammal, 49:287.
Watkins, W. A., and Schevill, W. E., 1971, Four hydrophone array
    for acoustic three-dimensional location, Technical Report
    71-60, Woods Hole Oceanographic Inst., Massachusetts.
Watkins, W. A., and Schevill, W. E., 1971, Underwater sounds of
    Monodon (Narwhal), Jour. Acoust. Soc. Amer., 49(2):595.
Watkins, W. A., and Schevill, W. E., 1972, Sound source location by
    arrival-times on a non-rigid three-dimensional hydrophone
    array, Deep-Sea Research, 19:691.
Watkins, W. A., and Schevill, W. E., 1974, Listening to Hawaiian
    spinner porpoises, Stenella cf. longirostris, with a three-
    dimensional hydrophone array, Jour. Mammal., 55:319.
Watkins, W. A., and Schevill, W. E., 1975, Sperm whales (Physeter
    catodon) react to pingers, Deep-Sea Research, 22:123.
Watkins, W. A., and Schevill, W. E., 1976, Right whale feeding and
    baleen rattle, Jour. Mammal., 57:58.
Watkins, W. A., and Schevill, W. E., 1977, Spatial distribution of
    Physeter catodon (sperm whales) underwater, Deep-Sea Research,
    24:693.
Watkins, W. A., and Schevill, W. E., 1977, Sperm whale codas, Jour.
    Acoust. Soc. Amer., 62:1485.
Watkins, W. A., Schevill, W. E., and Best, P. B., 1977, Underwater
    sounds of Cephalorhynchus heavisidii (Mammalia: Cetacea),
    Jour. Mammal., 58:316.
Watkins, W. A., Schevill, W. E., and Ray, G. C., 1971, Underwater
    sounds of Monodon (Narhwal), Jour. Acous. Soc. Amer., 49:595.
Wedmid, G., Litchfield, C., Ackman, R. G., and Sipos, J. C., 1971,
    Heterogeneity of lipid composition within the cephalic melon
    tissue of the pilot whale (Globicephala melaena), Jour. Amer.
    Oil Chem. Soc., 48:332.
Wedmid, G., Litchfield, C. Ackman, R. G., Sipos, J. C., Eaton, C.
    A., and Mitchell, E. D., 1973, Heterogeneity of lipid compo-
    sition within the cephalic melon tissue of the pilot whale
    (Globicephala melaena), Biochemica & Biophysica Acta, 326:439.
Wever, E. G., McCormick, J. G., Palin, J., and Ridgway, S. H., 1971,
    The cochlea of the dolphin Tursiops truncatus: general
    morphology, Proc. Nation. Acad. Sci., USA, 68:2381.
Wever, E. G., McCormick, J. G., Palin, J., and Ridgway, S. H.,
    1971, The cochlea of the dolphin Tursiops truncatus: the
    basilar membrane, Proc. Nation. Acad. Sci., USA, 68:2708.
Wever, E. G., McCormick, J. G., Palin, J., and Ridgway, S. H.,
    1971, The cochlea of the dolphin, Tursiops truncatus: hair
    cells and ganglion cells, Proc. Nation. Acad. Sci., USA,
    68:2908.

Wever, E. G., McCormick, J. G., Palin, J., and Ridgway, S. H.,
    1972, Cochlear structure in the dolphin Lagenorhynchus
    obliquidens, Proc. Nation. Acad. Sci., USA, 69:657.

White, M .J., Ljungblad, D., Norris, J. and Baron, K., 1977,
    Auditory thresholds of two Beluga whales, Proceedings
    (abstracts) Second Conference on the Biology of Marine
    Mammals, San Diego, Calif.

Whitmore, F. C., Jr., and Sanders, A. E., 1976, Review of the
    Oligocene Cetacea, Syst. Zool., 25(4):304.

Winn, H. E., Perkins, P. J., and Poulter, T. C., 1970, Sounds of
    the humpback whale, in: "Proceedings of the Seventh Annual
    Conference on Biological Sonar and Diving Mammals", Stanford
    Research Institute, Menlo Park, Calif.

Winn, H. E., Perkins, P. J., and Winn, L., 1970, Sounds and behavi-
    our of the northern bottle-nosed whale, in: "Proceedings of
    the Seventh Annual Conference on Biological Sonar and Diving
    Mammals", Menlo Park, Calif.

Wood, F. G., 1973, "Marine Mammals and Man", R. B. Luce, Inc.,
    New York.

Wood, F. G., 1978, The cetacean stranding phenomenon: An hypothe-
    sis, in: "Report on the Marine Mammal Stranding Workshop",
    N. T. I. S., Washington, D.C.

Yablokov, A. V., Bel'kovich, V. M., and Borisov, V. I., 1972,
    Whales and Dolphins, Parts I and II, Kity i Del'fini, Izd-vo
    Nauka, Moscow (English translation JPRS 62150-1).

Yanagisawa, K., Sato, O., Nomoto, M., Katsuki, Y., Ikeono, E.,
    Grinnel, A. D., and Bullock, T. H., 1966, Auditory evoked
    potentials from brain stem in cetaceans, Fed. Proc.
    25:464.

Yunker, M. P., and Herman, L. M., 1974, Discrimination of auditory
    temporal differences by the bottlenose dolphin and by the
    human, Jour. Acous. Soc. Amer., 56:1870.

Zaslavskiy, G. L., 1971, On directivity of sound emission in the
    Black Sea bottlenosed dolphin, Trudy Akusticheskogo Instituta,
    17:60.

Zaslavskiy, G. L., 1972, Study of echolocation signals of bottle-
    nose dolphin (Tursiops truncatus) two channel system of
    registration, Biofizika, 17:717.

Zaslavskiy, G. L., 1974, Experimental study of time and space
    structure of the dolphin's echolocating signals, candida-
    torial dissertation, Karadag, as cited in: Bel'kovich, V. M.
    and Dubrovskiy, N. A., 1976, "Sensory Bases of Cetacean
    Orientation", Nauka, Leningrad (English translation JPRS
    L/7157).

Zaslavskiy, G. L., Titov, A. A., and Lekomtsev, V. M., 1969,
    Investigation of hydrolocation capabilities of the Azov
    dolphin (Phocoena phocoena), Trudy Akusticheskogo Instituta,
    8:134.

Zaytseva, K. A., Akopian, A. I., and Morozov, V. P., 1975, Noise
    resistance of the dolphin auditory analyzer as a function of
    noise direction, Biofizika, 20(3):519.
Zlatoustova, L. V., and Nizova, A. B., 1971, An acoustical analysis
    of some whistles of the bottlenose dolphin (Tursiops
    truncatus Montagu), "Morfologiya i Ekologiya Morskikh
    Mlekopitayushchikh", V. Ye. Sokolov, ed., Izdatel'stvo
    Nauka, Moscow.

This bibliography was compiled and prepared by Paul E. Nachtigall.

The assistance of Vladimir Gurevich and Mike Fielding in the
acquisition of Russian references is gratefully acknowledged.
Jeff Haun and Earl Murchison graciously assisted in proofreading,
while Phyllis Johnson expertly typed the entire list.

Bibliography on echolocation in non-bat, non-cetacean species

1. Buchler, E. R. The use of echolocation by the wandering shrew Sorex vagrans. Anim. Behav., 24, 858-873, 1976.
2. Cranbrook, E. and Medway, L. Lack of ultrasonic frequencies in the calls of swiftlets. Ibis, 107, 258. 1965.
3. Eisenberg, J. F. and Gould, E. The behavior of Solenodon paradoxus in captivity with comments on the behavior of other insectivora. Zool. 51, 49, 1966.
4. Eisenberg, J. F. and Gould, E. The tenrecs: a study in mammalian behavior and evolution. Smithsonian Contributions to Zoology No. 27, 138 p.
5. Fenton, M. B. The role of echolocation in the evolution of bats. Am. Nat., 108, 386-388. 1974.
6. ----- Acuity of echolocation in Collocalia hirundinacea (Aves: Apodidae), with comments on the distribution of echolocating swiftlets and molossid bats. Biotropica, 7, 1-7. 1975.
7. Gans, C. and Maderson, P. F. A. Sound producing mechanisms in recent reptiles: review and comment. Amer. Zool., 13, 1195-1203. 1973.
8. Gehlback, F. R. and Walker, B. Acoustic behavior of the aquatic salamander, Siren intermedia. BioScience, 20, 1107-1108. 1970.
9. Griffin, D. R. and Suthers, R. A. Sensitivity of echolocation in cave swiftlets. Biol. Bull., 139, 495-501. 1970.
10. Griffin, D. R. and Buchler, E. Echolocation of extended surfaces. In: "Animal migration, navigation and homing." Ed. by K. Schmidt-Koenig, New York, Springer-Verlag, 1978. p. 201-08.
11. Grunwald, A. Untersuchungen zur orientiergun der weisszahnspitzmaus (Soricidae-Crocidurinae). Z. vergl. Physiol., 65, 191-217. 1969.
12. Harrisson, T. Onset of echolocation clicking in Collocalia swiftlets. Nature, 212, 530-531, 1966.
13. Holyoak, D. T. Undescribed land birds from the Cook Islands, Pacific Ocean. Bull. Br. Ornith. Club, 94, 145-150. 1974.
14. Hutterer, V. R. Beobachtungen zur geburt and jugendentwicklung der zwergspitzmaus, Sorex minutus L. (Soricidae - Insectivora). Z. Saugetierkunde, 41, 1-22. 1976.
15. Ilichev, V. D. Certain problems in studying the orientation of birds. U. S. Gov. Res. Dev. Rep. 70, 173. 1970.
16. Lore, R., Kam, B. and Newby, V. Visual and nonvisual depth avoidance in young and adult rats. J. Comp. and Physiol. Psychol., 65, 525-528. 1967.
17. Komarov, V. T. Underwater sounds of the muskrat, Ondatra zibethica and the water vole Arvicola terrestris. Zool. Zh., 55, 632-633. 1976.

18. Konishi, M. and Knudsen, E. I. The oilbird: hearing and
     echolocation. Sci. 204, 425-427. 1979.

19. Konstantinov, A.I., Akhmarova, N.I. and Golovina, S.S.
     The possibility of small rodents using ultrasonic location.
     In: "Echolocation in animals." Ed. by E. Sh. Airapetyants,
     A. I. Konstantinov, Springfield, Virginia, U.S. Department
     of Commerce, National Tech. Info. Serv., 1966. 390 pp.

20. Medway, L. The swiftlets (Collocalia) of Java, and their
     relationships. J. Bombay Nat. Hist. Soc., 59, 146-153. 1962.

21. -----Field characters as a guide to the specific relations
     of swiftlets. Proc. Linn. Soc. Lond., 177, 151-172. 1966.

22. -----The function of echonavigation among swiftlets.
     Anim. Behav., 15, 416-420. 1967.

23. -----Studies on the biology of the edible nest swiftlets
     of southeast Asia. Malay, Nat. J., 22, 57-63. 1969.

24. -----The nest of Collocalia v. vanikorensis, and
     taxonomic implications. Emu, 75, 154-155. 1975.

25. Medway, L. and Wells, D. R. Dark orientation by the giant
     swiftlet Collocalia gigas. Ibis, III, 609-11. 1969.

26. -----The Birds of Malay Peninsula, Vol. 5, London,
     Witherby. 1976.

27. Medway, L. and Pye, J. D. Echolocation and the systematics
     of swiftlets. In: "Evolutionary ecology." Ed. by
     B. Stonehouse and C. Perrins, Baltimore, Maryland,
     University Park Press, 1977. p. 225-238.

28. Pecotich, L. Grey swiftlets in the Tully River gorge and
     Chillagoe Caves. Sunbird, 5, 16-21. 1974.

29. Penny, M. The birds of Seychelles and the outlying
     islands. London, Collins. 1976.

30. Pernetta, J. C. Anatomical and behavioral specializations
     of shrews in relation to their diet. Can. J. Zool. 55,
     1442-1453. 1977.

31. Player, J. In search of Guianas oilbirds. Animals, 15,
     74-79. 1973.

32. Sato, Y. Visual sense organs of soricoidea. Zool. Mag.,
     81, 419. 1972.

33. Schwartzkopff, J. Auditory communication in lower animals
     role of auditory physiology. In: "Annual review of
     psychology", Vol. 28. Ed. by M. R. Rosenzweig and
     L. W. Porter, Palo Alto, California, Annual Reviews, Inc.,
     1977. p. 61-84.

34. Sergeev, V. E. Characteristics of the orientation of
     shrews in water. Ekologiya, 4, 87-90. 1973.

35. Simmons, J. A. Echolocation for perception of the
     environment. J. Acoust. Soc. Am., 58, S18. 1975.

36. Smythies, B. E. A note on the swiftlets Collocalia found
     in Burma. J. Bombay Nat. Hist. Soc., 72, 847-851. 1975.

37. Snow, D. W. Oilbirds, cave-living birds of South America.
     Stud. Speleol. 2, 257-264. 1975.

38. Tavolga, W. N. Acoustic obstacle detection in the sea catfish (Arius felis). In: "Sound reception in fish." Ed. by A. Schuijf and A. D. Hawkins, Amsterdam, Elsevier Scientific Publ. Co., 1976. p. 185-203.

39. ----- Acoustic orientation in the sea catfish, Galeichthys felis. In: "Orientation: sensory basis." Ed. by H. Adler, Ann. N.Y. Acad. Sci., 188, 80-97, 1971.

40. ----- Mechanisms for directional hearing in the sea catfish (Arius felis). J. Exp. Biol. 67, 97-115. 1977.

41. Thurow, G. R. and Gould, H. J. Sound production in a caecilian. Herpetologica, 33, 234-237. 1977.

42. Vaughan, T. A. Mammalogy. Philadelphia, W. B. Saunders Co., 1972. 463 pp.

43. Vogel, P. Comparative investigations of the mode of ontogenesis of domestic Soricidae Crocidura russula, Sorex araneus and Neomys fodiens. Rev. Suisse Zool. 79, 1201-1332. 1972.

Compiled by Edwin Gould

SYMPOSIUM PARTICIPANTS

ALCURI (Gustavo)
Laboratoire d'Acoustique Animale
E.P.H.E. - I.N.R.A. - C.N.R.Z.
78350 Jouy-en-Josas   France

AU (Whitlow W.L.)
Naval Ocean Systems Center
P.O. Box 997
Kailua, Hawaii 96734   U.S.A.

BEUTER (Karl J.)
Battelle-Institut e.V.
Am Römerhof 35   F.R.
Frankfurt/Main   Germany

BODANHAMER (Robert)
Department of Zoology
University of Texas
Austin, Tx. 78712   U.S.A.

BOWERS (Clark A.)
Naval Ocean Systems Center
Code 5111
San Diego, Ca. 92152   U.S.A.

BROWN (Patricia)
Department of Biology
University of California
Los Angeles, Ca. 90024   U.S.A.

BRUNS (Volkmar)
Fachbereich Zoologie
J.W. Goethe-Universität   F.R.
D-6000 Frankfurt/Main   Germany

BUCHLER (Edward)
Department of Zoology
University of Maryland
College Park, Md. 10742   U.S.A.

BULLOCK (Theodore H.)   Department of Neurosciences
                             School of Medicine
                             University of California
                             San Diego, Ca. 92093   U.S.A.

BUSNEL (René-Guy)   Laboratoire d'Acoustique Animale
                             E.P.H.E. - I.N.R.A. - C.N.R.Z.
                             78350 Jouy-en-Josas   France

CHASE (Julia)   Biology Department
                             Columbia University
                             Barnard College
                             New York, N.Y. 10027   U.S.A.

DECOUVELAERE (Martine)   Ecole Nationale Supérieure
                             des Télécommunications
                             46, rue Barrault
                             75634 Paris Cedex 13   France

DIERCKS (K. Jerome)   Applied Research Laboratories
                             University of Texas
                             P.O. Box 8029
                             Austin, Tx. 78712   U.S.A.

DO (Manh Anh)   Radio Engineering Limited
                             P.O. Box 764
                             Dunedin   New Zealand

DZIEDZIC (Albin)   Laboratoire d'Acoustique Animale
                             E.P.H.E. - I.N.R.A. - C.N.R.Z.
                             78350 Jouy-en-Josas   France

ESCUDIE (Bernard)   Laboratoire du Traitement du
                             Signal - I.C.P.I.
                             25, rue du Plat
                             69288 Lyon   France

EVANS (William E.)   Hubbs-Sea World Research Institute
                             1700 South Shores Drive
                             San Diego, Ca. 92109   U.S.A.

FENG (Albert S.)   Department of Physiology and
                             Biophysics
                             University of Illinois
                             Urbana, Il. 61801   U.S.A.

FENTON (M. Brock)   Department of Biology
                             Carleton University
                             Ottawa, KIS 5B6   Canada

FIEDLER (Joachim)   Fachbereich Biologie-Zoologie
                           J.W. Goethe Universität   F.R.
                           D-6000 Frankfurt/Main   Germany

FISH (James F.)   Sonatech. Inc.
                           700 Francis Botello Road
                           Goleta, Ca. 93017   U.S.A.

FLEISCHER (Gerald)   c/o Umweltbundesant
                           Bismarckplatz 1   F.R.
                           D-1000 Berlin 33   Germany

FLOYD (Robert W.)   Naval Oceans Systems Center
                           P.O. Box 997
                           Kailua, Hawaii 96744   U.S.A.

FULLARD (James)   Department of Biology
                           Carleton University
                           Ottawa K1S 5B6   Canada

Von GIERKE (Henning E.)   Biodynamics and Bioengineering
                           Division
                           Aerospace Medical Research
                           Laboratory
                           Wright Patterson Air Force Base
                           Dayton, Oh. 45433   U.S.A.

GOULD (Edwin)   Johns Hopkins University
                           School of Hygiene and Public Health
                           615 North Wolfe Street
                           Baltimore, Md. 21205   U.S.A.

GREGUSS (Pal)   Applied Biophysics Laboratory
                           Technical University Budapest
                           H-1111 Budapest   Hungary

GRIFFIN (Donald R.)   Rockefeller University
                           New York, N.Y. 10021   U.S.A.

GRINNELL (Alan D.)   Department of Biology and
                           Physiology
                           University of California
                           Los Angeles, Ca. 90024   U.S.A.

GUREVICH (Vladimir Sol)   Hubbs-Sea World Research Institute
                           1700 South Shores Road
                           San Diego, Ca. 92109   U.S.A.

HALLS (Justin A.T.)   Department of Zoology and
                           Comparative Physiology
                           Queen Mary College
                           Mile End Road
                           London El 4NS   England

HENSON (O'Dell W.)   Department of Anatomy
                           School of Medicine
                           University of North Carolina
                           Chapel Hill, N.C. 27514   U.S.A.

JEN (Philip H.S.)   Division of Biological Sciences
                           University of Missouri
                           Columbia, Mo. 65201   U.S.A.

JOHNSON (C. Scott)   Naval Ocean System Center
                           Code 5102
                           San Diego, Ca. 92152   U.S.A.

JOHNSON (Richard A.)   Naval Ocean Systems Center
                           Code 5122
                           San Diego, Ca. 92152   U.S.A.

KAY (Leslie)   Department of Electrical Engineering
                           University of Canterbury
                           Christchurch   New Zealand

KICK (Shelley)   Department of Psychology
                           Washington University
                           St. Louis, Mo. 63130   U.S.A.

LEVY (Jean-Claude)   Bureau de Psychologie Appliquée
                           de la Direction du Personnel
                           Militaire de la Marine
                           3, Avenue Octave Gréard
                           75007 Paris   France

LONG (Glenis R.)   Central Institute for the Deaf
                           818 South Euclid
                           St. Louis, Mo. 63110   U.S.A.

Mc.CORMICK (James G.)   Department of Otolaryngology
                           Bowman Gray School of Medicine
                           Wake Forest University
                           Winston-Salem, N.C. 27103   U.S.A.

McKAY (R. Stuart)   Biology Department
                           Boston University
                           Boston, Mass. 02215   U.S.A.

MARCUS (Stanley R.)	Naval Sea Systems Command (Sea 038) Washington, D.C. 20363	U.S.A.
MILLER (Lee A.)	Department of Psychology Stanford University Stanford, Ca. 94305	U.S.A.
MOORE (Patrick W.B.)	SEACO, Inc. 146 Hekili St. Kailua, Hawaii 96734	U.S.A.
MURCHISON (Earl A.)	Naval Ocean Systems Center P.O. Box 997 Kailua, Hawaii 96734	U.S.A.
NACHTIGALL (Paul)	Naval Ocean Systems Center P.O. Box 997 Kailua, Hawaii 96734	U.S.A.
NEUWEILER (Gerhard)	Fachbereich Biologie-Zoologie Zoologisches Institut J.W. Goethe-Universität D-6000 Frankfurt/Main	F.R. Germany
NORRIS (Kenneth S.)	Coastal Marine Laboratory Environmental Studies University of California Santa Cruz, Ca. 95064	U.S.A.
O'NEILL (William E.)	Department of Biology Washington University St. Louis, Mo. 63130	U.S.A.
OSTWALD (Joachim)	Fachbereich Biologie-Zoologie Philipps-Universität Marburg Lahnberge 3550 Marburg/Lahn	F.R. Germany
PENNER (Ralph H.)	Naval Ocean Systems Center P.O. Box 997 Kailua, Hawaii 96734	U.S.A.
POLLACK (George)	Department of Zoology University of Texas Austin, Tx. 78712	U.S.A.
PORTER (Homer O.)	Biosciences Department Naval Ocean Systems Center Code 51 San Diego, Ca. 92152	U.S.A.

POWELL (William A.)   Naval Ocean Systems Center
                             Code 05
                             San Diego, Ca. 92152   U.S.A.

PYE (Ade)   Institute of Laryngology and
                             Otology
                             330, Gray's Inn Road
                             London WCIX 8EE   England

PYE (J. David)   Department of Zoology and
                             Comparative Physiology
                             Queen Mary College
                             Mile End Road
                             London E1 4NS   England

RADTKE (Susanne)   Fachbereich Biologie-Zoologie
                             J.W. Goethe-Universität   F.R.
                             D-6000 Frankfurt/Main   Germany

RIDGWAY (Sam H.)   Naval Ocean Systems Center
                             Code 5103 Seaside
                             San Diego, Ca. 92152   U.S.A.

RÜBSAMEN (Rudolf)   Fachbereich Biologie-Zoologie
                             J.W. Goethe-Universität   F.R.
                             D-6000 Frankfurt/Main   Germany

SCHLEGEL (Peter)   Fachbereich Biologie-Zoologie
                             J.W. Goethe Universität   F.R.
                             D-6000 Frankfurt/Main   Germany

SCHNITZLER (Hans-Ulrich)   Fachbereich Biologie-Zoologie
                             Philipps-Universität
                             Lahnberge   F.R.
                             3550 Marburg/Lahn   Germany

SCHULLER (Gerd)   Fachbereich Biologie-Zoologie
                             J.W. Goethe-Universität   F.R.
                             D-6000 Frankfurt/Main   Germany

SCHUSTERMAN (Ron)   Biology Department
                             California State University
                             Hayward, Ca. 94542   U.S.A.

SCHWEIZER (Hermann)   Fachbereich Biologie-Zoologie
                             J.W. Goethe-Universität   F.R.
                             D-6000 Frankfurt/Main   Germany

SIMMONS (James A.)   Departments of Psychology and
                          Biology
                          Washington University
                          St. Louis, Mo. 63130   U.S.A.

SUGA (Nobuo)   Department of Biology
                          Washington University
                          St. Louis, Mo. 63130   U.S.A.

SUTHERS (Roderick A.)   Medical Sciences Program -
                          Myers Hall
                          Indiana University
                          School of Medicine
                          Bloomington, Ind. 47401   U.S.A.

STONE (Harris B.)   Office of Chief of Naval
                          Operations (OP 987)
                          Washington, D.C. 20350   U.S.A.

TIPPER (Ronald C.)   Office of Naval Research
                          Oceanic Biology Branch
                          Code 484 - NSTL Station
                          Bay St. Louis, Miss. 39529   U.S.A.

VATER (Marianne)   Fachbereich Biology
                          J.W. Goethe-Universität   F.R.
                          D-6000 Frankfurt/Main   Germany

WATKINS (William A.)   Woods Hole Oceanographic Institute
                          Woods Hole, Mass. 02543   U.S.A.

WOOD (Forrest G.)   Biosciences Department
                          Naval Ocean Systems Center
                          Code 5104
                          San Diego, Ca. 92152   U.S.A.

ZOOK (John M.)   Department of Psychology
                          Duke University
                          Durham, N.C. 27706   U.S.A.

OBSERVERS

HARNISCHEFEGER (Günther)   Fachbereich Biologie-Zoologie
                           J.W. Goethe-Universität   F.R.
                           D-6000 Frankfurt/Main   Germany

HERVIEU (René)   20, rue du Calvaire
                           44000 Nantes   France

MENNE (Dieter)   Fachbereich Biologie
                           Philipps-Universität
                           Lahnberge   F.R.
                           3550 Marburg/Lahn   Germany

REISS (Diana)   Laboratoire d'Acoustique Animale
                           E.P.H.E. - I.N.R.A. - C.N.R.Z.
                           78350 Jouy-en-Josas   France

WHEELER (Leslie A.)   Laboratoire d'Acoustique Animale
                           E.P.H.E. - I.N.R.A. - C.N.R.Z.
                           78350 Jouy-en-Josas   France

ABSENT CONTRIBUTORS

ALTES (Richard A.)   Orincon Corporation
                           3366 N. Torrey Pines Ct.
                           La Jolla, Ca. 92037   U.S.A.

POPPER (Arthur N.)   Department of Anatomy
                           Georgetown University
                           Washington, D.C. 20007   U.S.A.

INDEXES

- Author Index

- Species Index

- Subject Index

Compiled by L. Wheeler

Author Index

Underlined numbers refer to pages on which the complete references are listed.

Species Index